Symplectic and Poisson Geometry on Loop Spaces of Smooth Manifolds and Integrable Equations

O.I. Mokhov

Cambridge Scientific Publishers

Reviews in Mathematics and Mathematical Physics
Volume 13, pp 1-222: 2009

Printed in UK

to my mother,

Maya N. Mokhova (Khlebnikova)

Symplectic and Poisson Geometry on Loop Spaces of Smooth Manifolds, and Integrable Equations

O. I. Mokhov

Centre for Nonlinear Studies,
L.D.Landau Institute for Theoretical Physics,
Russian Academy of Sciences,
Kosygina, 2,
Moscow, GSP-1, 117940 Russia,
e-mail: mokhov@landau.ac.ru,
mokhov@mi.ras.ru

ABSTRACT

This review is devoted to the differential-geometric theory of homogeneous forms and other different homogeneous structures (mainly, Poisson and symplectic ones) on loop spaces of smooth manifolds, their natural generalizations and applications in mathematical physics and field theory.

We study complexes of homogeneous forms on loop spaces of smooth manifolds and their cohomology groups, the differential geometry of homogeneous (local and non-local) symplectic and Poisson structures and their applications to Lagrangian and bi-Lagrangian systems, non-linear sigma-models, the equations of associativity in two-dimensional topological field theory, the Monge-Ampère equations, the Heisenberg ferromagnets, the Korteweg–de Vries equation, the Chaplygin equations, systems of hydrodynamic type, the Whitham equations, the N-wave equations, and other non-linear systems of geometry, hydrodynamics, gas dynamics, field theory, mathematical and theoretical physics.

An up-to-date review of the theory of Poisson brackets of hydrodynamic type is presented. We consider in detail one-dimensional and multidimensional, local and non-local, degenerate and non-degenerate, homogeneous and non-homogeneous, Poisson structures of hydrodynamic type, their higher-order generalizations, and the corresponding non-linear Hamiltonian and multi-Hamiltonian systems.

We state the theory of the equations of associativity in two-dimensional topological field theory as non-diagonalizable integrable homogeneous systems of hydrodynamic type. In particular, we describe Hamiltonian, bi-Hamiltonian and symplectic representations and conservation laws of the equations of associativity, and also finite dimensional reductions of the equations of associativity to the sets of stationary points of integrals. We consider the general important construction of restriction of arbitrary evolutionary system to the set of stationary points of its integrals, construct explicit canonical Hamiltonian representations of the corresponding reductions and study various applications of this construction.

Contents

1 Introduction

This survey paper is devoted to Hamiltonian geometry of integrable partial differential equations and the theory of infinite-dimensional symplectic and Poisson structures. Infinite-dimensional symplectic and Poisson geometry is closely connected with modern mathematical physics, field theory, and the theory of integrable systems of partial differential equations. In particular, structures studied in the present paper, namely, symplectic and Poisson structures of special differential-geometric type on loop spaces of manifolds, generate Hamiltonian representations for a number of important non-linear systems of hydrodynamics, gas dynamics, mathematical and theoretical physics, geometry and field theory, for example, such as systems of hydrodynamic type (these systems arise not only in Euler hydrodynamics and gas dynamics but also, in particular, if we apply the Whitham averaging procedure to the equations of the soliton theory [188], [87], [54], [34], [36]), equations of associativity in topological field theory [189], [190], [17], [29], [30] (these equations also play one of the key roles in the theory of Gromov–Witten invariants, which is being developed at present, the theory of quantum cohomology, and certain classic enumerative problems of algebraic geometry [176], [78], [79]), non-linear sigma-models, the Korteweg–de Vries equation, the Heisenberg ferromagnets, the Monge–Ampère equations, the N-wave equations, the Chaplygin equations, and many others. The presence of the Hamiltonian representations enables us to apply differential-geometric, symplectic, Hamiltonian, and bi-Hamiltonian methods effectively to the investigation of these systems, to study integrable classes of these systems, and to elucidate the differential-geometric, symplectic, and bi-Hamiltonian nature of the integrable cases.

The study of a special class of field-theoretical Poisson structures, namely, Poisson brackets of the special differential-geometric type, is initiated in the works of Dubrovin and Novikov [34], [35], where the general Hamiltonian formalism of systems of hydrodynamic type is developed in connection with the investigation of the Whitham equations describing an evolution of slowly modulated multiphase solutions of systems of partial differential equations. The Poisson structures introduced and studied by Dubrovin and Novikov (local Poisson structures of hydrodynamic type) are

generated by arbitrary flat pseudo-Riemannian metrics and can
be always reduced to a constant form by local changes of coordinates in coordinate neighbourhoods on the corresponding manifolds. On the basis of this Hamiltonian approach, Tsarev [181]
developed the differential-geometric theory (the generalized hodograph method) of integrating for diagonalizable Hamiltonian (and
also semi-Hamiltonian) systems of hydrodynamic type, in particular, the Whitham equations obtained as a result of averaging for
the Korteweg–de Vries equation.

In the works of the present author and Ferapontov ([151], see
also [121], [39], [152]) a non-local generalization of the Poisson
structures of Dubrovin and Novikov is found; this is generated
by arbitrary pseudo-Riemannian metrics of constant Riemannian
curvature (in this case, the results of Dubrovin and Novikov correspond to metrics of zero Riemannian curvature) and also plays
a very important role in the theory of systems of hydrodynamic
type, in particular, in the theory of the Whitham equations. Further non-local differential-geometric generalizations of the Poisson
structures of hydrodynamic type were studied by Ferapontov in
[39].

In the Hamiltonian theory of multidimensional systems of hydrodynamic type, there arise new differential-geometric and algebraic problems on the description of tensor obstructions to reducing
local multidimensional Poisson structures of hydrodynamic type to
a constant form and on the classification of these structures, which
is equivalent to the classification of infinite-dimensional Lie algebras of a special form (Lie algebras of hydrodynamic type) [35], [36],
[105], [115], [127] (the complete set of tensor relations for the obstructions in the general case and the complete explicit algebraic
classification in the case where the number of components is small
$(1 \leq N \leq 4)$ are obtained by the present author in the works
[105], [115]).

The study of non-homogeneous Poisson structures of hydrodynamic type, which have important applications, in particular, in
the theories of Heisenberg ferromagnets, the Korteweg–de Vries
equation, N-wave equations, and other non-homogeneous systems
of hydrodynamic type, leads to a special class of Kac–Moody algebras and the theory of Killing–Poisson bivectors on manifolds of
constant Riemannian curvature. The theory of the Killing–Poisson

bivectors is constructed in [115], [127], [152].

General local homogeneous Poisson structures of arbitrary orders are considered for the first time by Dubrovin and Novikov in [35], where the problem of their classification is posed. At present this problem is far from being completely solved. Only homogeneous Poisson structures of zero order (Darboux), the first order (Dubrovin and Novikov [34]), and the second order (Potemin [172], [173], Doyle [26]) are completely classified. And only partial classification results have been obtained for homogeneous Poisson structures of the third order (Potemin [172], Doyle [26]). The homogeneous Poisson structures of the third order play a very important role (in particular, they take part in bi-Hamiltonian representations) in the theories of Monge–Ampère equations (Mokhov and Nutku [154], see also [127]) and equations of associativity in two-dimensional topological field theory (Ferapontov, Galvão, Mokhov, and Nutku [50], see also [127]).

General symplectic structures of differential-geometric type, homogeneous symplectic operators of arbitrary orders, and homogeneous symplectic and presymplectic forms on loop spaces of smooth manifolds are introduced and studied in the works of the present author [109], [110], [115], [117], [124], [126], [127]. Their consideration has natural motivations connected, in particular, with non-linear sigma-models, string theory, and the corresponding geometry of loop spaces. As the present author showed in [109], the motion equations of two-dimensional non-linear sigma-models with torsion (such sigma-models define, in particular, the classic action for boson strings) possess Hamiltonian representations generated by homogeneous symplectic structures of the first order on the loop spaces of pseudo-Riemannian manifolds, that is, structures of the form

$$\omega(\xi, \eta)[\gamma] = \int_\gamma \langle \xi, \nabla_{\dot\gamma} \eta \rangle, \qquad (1.1)$$

where γ is a loop on a manifold M, $\gamma\colon S^1 \to M$, $x \in S^1$, $\gamma(x)$ is a point on M, $\gamma(x) \in M$, $T_{\gamma(x)}M$ is the tangent space of the manifold M at the point $\gamma(x)$, ξ and η are arbitrary smooth vector fields on the manifold M defined along the loop γ, $\xi(\gamma(x))$, $\eta(\gamma(x)) \in T_{\gamma(x)}M$, $\langle \xi, \eta \rangle = g_{ij}(\gamma(x))\xi^i(\gamma(x))\eta^j(\gamma(x))$ is the scalar product on $T_{\gamma(x)}M$ defined by a pseudo-Riemannian metric g_{ij} on the manifold M, $\dot\gamma$ is the velocity vector field of the loop γ, and $\nabla_{\dot\gamma}$ is a covariant differentiation along the loop γ on M. In the works of the present author [109], [110], the problem of the description of homogeneous symplectic (presymplectic) forms on loop spaces of smooth manifolds is completely solved for symplectic forms of the first and second orders (symplectic forms of zero order are classified by the classic Darboux theorem of finite-dimensional symplectic geometry). In particular, symplectic structures of the first order (1.1) are specified by affine connections on an arbitrary pseudo-Riemannian manifold such that these connections are compatible with the corresponding pseudo-Riemannian metric of the manifold and their torsion tensors define closed 3-forms on this pseudo-Riemannian manifold [109], [124]. Symplectic structures of the second order are generated precisely by symplectic connections on almost symplectic manifolds [110], [124].

The interpretation of the homogeneous structures (arbitrary homogeneous symplectic operators and arbitrary homogeneous Poisson brackets) given by homogeneous differential operators as natural differential-geometric symplectic (presymplectic) forms and Poisson brackets on loop spaces of smooth manifolds was proposed for the first time by the present author in [109], [110]. Such forms and brackets are expressed by means of invariant operators of covariant differentiations along loops (see, for example, formula (1.1)).

The question of the connection of our symplectic structures (1.1) with finite-dimensional symplectic structures, that is, the question of reductions of the form (1.1) to finite-dimensional manifolds, is considered for the first time and studied in the present author's paper [109]. For N-dimensional Riemannian manifolds (M, g_{ij}) whose geodesics are periodic and are of equal length, the manifold of geodesics CM is a $(2N - 2)$-dimensional symplectic manifold. The tangent space $T_\gamma CM$ of the manifold of geodesics CM at the point γ is isomorphic to the space of normal Jacobian fields along the geodesic γ on the manifold M. In the present author's paper [109] the symplectic structures (1.1) on the loop spaces ΩM of such special Riemannian manifolds M were considered, and the restrictions of these symplectic structures to the finite-dimensional manifolds CM of closed geodesics on the manifolds M were studied. In particular, in [109] it is proved that the restriction of the symplectic structure (1.1) defined on the loop space

ΩM of the manifold M by the Levi-Civita connection to the finite-dimensional subspace of normal Jacobian fields along the geodesic γ coincides with the Reeb form, which is a closed non-degenerate 2-form defining a finite-dimensional symplectic structure on CM. In the present author's works [114], [116], general homogeneous k-forms of arbitrary orders on loop spaces of smooth manifolds were defined and studied, the complexes $(\Omega_{[m]}, d)$ of homogeneous forms of arbitrary order m were constructed, and the cohomology groups of these complexes (homogeneous cohomology groups of order m of loop spaces of smooth manifolds) were found for a number of cases.

In this paper, particular attention is given to symplectic and Poisson structures and also k-forms that are invariant with respect to the action of the group $\mathrm{Diff}^+(S^1)$ of orientation-preserving diffeomorphisms of the circle S^1, that is, to structures on loop spaces that do not depend on parametrizations of the loops. This is important, in particular, from the viewpoint of applications in the theory of closed boson strings in curved N-dimensional space-time M with the metric (gravitational field) g_{ij}. The configuration space of these closed boson strings is the loop space ΩM of the pseudo-Riemannian manifold (M, g_{ij}) (the requirement of invariance or, in other words, independence on the parametrizations of the loops is necessary in the theory of boson strings for physically sensible objects on the configuration space ΩM).

Moreover, compatible symplectic and Poisson structures, generating, as Magri discovered [88], [89], integrable bi-Hamiltonian non-linear systems, are of particular interest. The study of the compatibility condition for symplectic and Poisson structures on loop spaces of manifolds leads to new interesting and non-trivial differential-geometric problems, the solution of which makes possible essential progress in the investigation of integrability properties of the corresponding non-linear systems of partial differential equations. It is necessary to study the compatibility conditions for natural classes of Poisson and symplectic structures. For the first time the problem of such type was considered by the present author in [101], [103], where all compatible local scalar Poisson structures of the first and the third orders, that is, all compatible Poisson brackets determined by arbitrary scalar ordinary differential operators of the first and the third orders, were completely described and classified. The class of compatible Poisson brackets found in the works [101], [103] is a natural generalization of the well-known pair of compatible Poisson brackets generating the Korteweg–de Vries equation, namely, the bracket of Gardner, Zakharov, and Faddeev (the Poisson bracket of the first order [63], [196]) and the bracket of Magri (the Poisson bracket of the third order [88]). Afterwards Cooke developed this approach in [16], where compatible local scalar Poisson brackets up to the fifth order were considered.

The important problem of classification or effective description of compatible Poisson structures of hydrodynamic type (local and non-local) is very interesting and highly non-trivial (see [29], [46], [47], [56], [118], [119], [128]–[150]). The equations for non-singular pairs of compatible non-degenerate local Poisson structures of hydrodynamic type (or, in other words, non-singular pairs of compatible flat metrics) have been integrated by the present author in the works [132]–[137] (see also [47]). As is shown by the present author in [118], [128]–[130], even in the two-component case, this problem is non-trivial and equivalent to integrating a curious integrable non-diagonalizable four-component homogeneous system of hydrodynamic type possessing only two Riemann invariants. For this four-component system, in particular, some integrable two-component and three-component reductions have been constructed [118], [128]–[130]. Some of these reductions are generated by Riemann invariants of the system. But there is also an integrable reduction connected with the associativity equations. And in the general N-component case there are always natural reductions to special solutions of the equations of associativity [29], [31], [118], [119], [128]–[130] (a deep connection between certain special classes of compatible local Poisson structures of hydrodynamic type and the theory of Frobenius manifolds was discovered by Dubrovin [29], [31]).

The problem of description of all non-singular pairs of compatible local Poisson structures of hydrodynamiic type has been solved in the present author's works [132]–[137] (see also [47]). This problem is equivalent, in particular, to the problem of local classification of quasi-Frobenius manifolds. In the works [132]–[137], the nonlinear partial differential equations describing all such compatible Poisson structures are obtained, and the integration procedure by the method of the inverse scattering problem for these equations is found. This integration procedure is based on reducing our equations in special "diagonal" local coordinates to a special non-linear differential reduction of the classical Lamé equations describing all n-orthogonal curvilinear coordinate systems in a flat n-dimensional space and on using the method of differential reductions proposed by Zakharov in the theory of the N-wave equations in [195]. The procedure reduces the problem to solving linear differential and linear integral equations. In the works [139], [140], bi-Hamiltonian integrable hierarchies of hydrodynamic type generated by arbitrary pairs of compatible local Poisson structures of hydrodynamic type are explicitly constructed. In the present author's works, all these results are generalized to compatible non-local Poisson brackets of hydrodynamic type generated by metrics of constant Riemannian curvature [141]–[144] and to compatible general non-local Poisson brackets of hydrodynamic type [145]–[149]. In particular, nonlinear equations describing all non-singular pairs of compatible general non-local Poisson brackets of hydrodynamic type are obtained, a Lax pair with spectral parameter for these equations is found, and the integrability of these equations by the method of the inverse scattering problem is proved. In addition, integrable reductions of the Lamé equations are constructed. These reductions are connected with arbitrary pairs of compatible Poisson brackets of hydrodynamic type one of which is local. And also integrable reductions of nonlinear equations describing all n-orthogonal curvilinear coordinate systems in an n-dimensional space of constant Riemannian curvature are constructed. In this series of papers, the theory of compatible Riemannian and pseudo-Riemannian metrics, which is motivated by the theory of compatible Poisson structures of hydrodynamic type, is constructed. It is also proved that two arbitrary Dubrovin–Novikov Hamiltonian operators are compatible if and only if one of these operators is the Lie derivative of the other operator along a certain vector field [138], [140]. Moreover, integrable N-parameter deformations of quasi-Frobenius algebras generated by arbitrary compatible $N \times N$-metrics of constant Riemannian curvature are constructed in [150].

Note also that, generally speaking, in infinite-dimensional symplectic and Poisson geometry there is no analogue of the classic general Darboux theorem of finite-dimensional symplectic geometry on reducing arbitrary finite-dimensional symplectic and arbitrary finite-dimensional non-degenerate Poisson structures to a canonical constant form. Therefore the study of the possibility of reducing infinite-dimensional symplectic and Poisson structures to a canonical constant form (in this case, we can effectively apply, in particular, the methods of perturbation theory for investigations of the corresponding Hamiltonian system of partial differential equations) is a very important current problem.

2 Differential geometry of symplectic structures on loop spaces of smooth manifolds

2.1 Symplectic and Poisson structures on loop spaces of smooth manifolds. Basic definitions

2.1.1 Complex of skew-symmetric forms on loop spaces of smooth manifolds

Let M be a smooth N-dimensional manifold with local coordinates $u^1, \ldots, u^N$. A *loop* γ on the manifold M is an arbitrary smooth parametrized map of the circle S^1 into M, $\gamma\colon S^1 \to M$, $\gamma(x) = \{u^i(x), \, 1 \le i \le N\}$, $x \in S^1$. In this paper by the *loop space* ΩM of the manifold M we mean the space of all smooth parametrized maps γ of the circle S^1 into M. The tangent space $T_\gamma \Omega M$ of the loop space ΩM at the point (the loop) γ consists of all smooth vector fields $\xi = \{\xi^i, \, 1 \le i \le N\}$, defined along the loop γ, with $\xi(\gamma(x)) \in T_{\gamma(x)} M$, for all $x \in S^1$, where $T_{\gamma(x)} M$ is the tangent space of the manifold M at the point $\gamma(x)$.

Let us introduce the notation A^k, $k \ge 0$, for the linear space of all skew-symmetric k-forms $\omega^k(\xi_1, \ldots, \xi_k)[\gamma]$ on the loop space ΩM, that is, the space of multilinear maps

$$\omega^k[\gamma]\colon \underbrace{T_\gamma \Omega M \times \cdots \times T_\gamma \Omega M}_{k \text{ copies}} \to \mathbf{R},$$

defined at all points (loops) $\gamma \in \Omega M$ and satisfying the condition of skew-symmetry

$$\omega^k(\xi_{\sigma(1)}, \ldots, \xi_{\sigma(k)})[\gamma] = \operatorname{sgn} \sigma \, \omega^k(\xi_1, \ldots, \xi_k)[\gamma],$$

where σ is an arbitrary permutation of k elements,

$$\operatorname{sgn} \sigma = \prod_{1 \le i < j \le k} \frac{\sigma(j) - \sigma(i)}{j - i} = \pm 1$$

is the sign of the permutation σ, $\xi_i \in T_\gamma \Omega M$, $\xi_i = (\xi_i^1, \ldots, \xi_i^N)$, $1 \le i \le k$. Note that the linear space A^0 (the space of 0-forms on ΩM)

is defined as the space of functionals ω^0 on the loop space ΩM:

$$\omega^0 \colon \Omega M \to \mathbf{R}.$$

The *differential* d on the linear space A^k, $k \geq 0$, is defined by the following formula (this is a natural generalization of the usual differential for k-forms on the Lie algebra of vector fields on a manifold):

$$(d\omega)(\xi_1, \ldots, \xi_{k+1})[\gamma] = \qquad\qquad (2.1)$$

$$= \sum_{i=1}^{k+1} \sum_{s=1}^{N} (-1)^{i-1} \int_{S^1} \xi_i^s \frac{\delta\omega(\xi_1, \ldots, \widehat{\xi_i}, \ldots, \xi_{k+1})[\gamma]}{\delta u^s(x)} \, dx +$$

$$+ \sum_{i<j} (-1)^{i+j} \omega([\xi_i, \xi_j], \xi_1, \ldots, \widehat{\xi_i}, \ldots, \widehat{\xi_j}, \ldots, \xi_{k+1})[\gamma],$$

where, in the general case, the commutator of vector fields is defined by the Gâteaux derivatives of one of the vector fields along another:

$$[\xi, \eta] = \eta_u' \xi - \xi_u' \eta, \qquad\qquad (2.2)$$

$$\xi_u' \eta = \left(\frac{d}{d\varepsilon} \xi[u + \varepsilon\eta] \right)\Big|_{\varepsilon=0}. \qquad\qquad (2.3)$$

For arbitrary smooth vector fields of local type

$$\eta = \eta(x, u, u_x, u_{xx}, \ldots, u_{(p)}) \quad \text{and} \quad \zeta = \zeta(x, u, u_x, u_{xx}, \ldots, u_{(r)})$$

this definition results in the formula

$$[\eta, \zeta]^j = \eta_{(s)}^i \frac{\partial \zeta^j}{\partial u_{(s)}^i} - \zeta_{(s)}^i \frac{\partial \eta^j}{\partial u_{(s)}^i}, \qquad\qquad (2.4)$$

where $f_{(s)} = d^s f/dx^s$, $d/dx = \partial/\partial x + u_x \partial/\partial u + u_{xx}\partial/\partial u_x + \cdots$, is the total sth derivative with respect to the independent variable x.

The differential d is a linear map

$$d \colon A^k \to A^{k+1}, \quad k \geq 0,$$

with $d^2 = 0$. Thus we have the following complex of skew-symmetric forms on the loop space ΩM:

$$0 \xrightarrow{d} A^0 \xrightarrow{d} A^1 \xrightarrow{d} A^2 \xrightarrow{d} \cdots. \qquad\qquad (2.5)$$

2.1.2   Symplectic structures

In the present paper we are interested, first of all, in closed skew-symmetric 2-forms $\omega(\xi, \eta)[\gamma]$ on the loop space ΩM, that is, for any loop γ and arbitrary vector fields ξ, η, ζ, defined along the loop, the following conditions must be satisfied:

$$\omega(\xi, \eta)[\gamma] = -\omega(\eta, \xi)[\gamma] \quad (\text{skew} - \text{symmetry}), \qquad (2.6)$$

$$(d\omega)(\xi, \eta, \zeta)[\gamma] = 0 \quad (\text{closedness}). \qquad (2.7)$$

According to (2.1) the differential of skew-symmetric 2-forms $\omega(\xi, \eta)[\gamma]$ is defined by the formula

$$(d\omega)(\xi, \eta, \zeta)[\gamma] =$$
$$= \sum_{(\xi, \eta, \zeta)} \left\{ \int_{S^1} \xi^i \frac{\delta\omega(\eta, \zeta)[\gamma]}{\delta u^i(x)} \, dx - \omega([\xi, \eta], \zeta)[\gamma] \right\}, \qquad (2.8)$$

where $\xi, \eta, \zeta \in T_\gamma \Omega M$, and here and everywhere in the present paper $\sum_{(\xi, \eta, \zeta)}$ means summation over all cyclic permutations of the elements (ξ, η, ζ).

Let us consider a bilinear form $\omega(\xi, \eta)[\gamma]$ on the loop space ΩM:

$$\omega(\xi, \eta)[\gamma] = \int_{S^1} \xi^i M_{ij} \eta^j \, dx, \qquad (2.9)$$

where $\xi, \eta \in T_\gamma \Omega M$, $\xi = (\xi^1, \ldots, \xi^N)$, $\eta = (\eta^1, \ldots, \eta^N)$, and $M = (M_{ij})$ is a linear operator depending on the loop $\gamma \in \Omega M$:

$$M[\gamma]: \ T_\gamma \Omega M \to T_\gamma^* \Omega M.$$

Definition 2.1 The operator M_{ij} is called *symplectic* and the 2-form ω is called a *presymplectic form* (or *presymplectic structure*) on the loop space ΩM if conditions (2.6) and (2.7) are fulfilled.

Remark 2.1 In the infinite-dimensional case presymplectic structures arising in the theory of integrable equations of mathematical physics and field theory are usually called symplectic even if the 2-form ω is degenerate on ΩM, that is, there exists a non-zero smooth vector field ξ along a loop γ such that $\omega(\xi, \eta)[\gamma] = 0$ for all

$\eta \in T_\gamma \Omega M$. In the present paper we shall also adhere to this terminology. A detailed introduction to the general theory of symplectic and Poisson structures on infinite-dimensional functional spaces in the framework of formal calculus of variations, generally accepted terminology, basic notions, in particular, general formal definitions of the complex of k-forms and the differential d, and also general symplectic and Hamiltonian conditions for differential operators in formal calculus of variations and all necessary references can be found, for example, in [20], [23], and also in [67], [69], [70], [24], [60], [55].

Under a local change of coordinates $u^i = u^i(\tilde{u})$ on the manifold M a symplectic operator $M_{ij}[u(x)]$ is transformed as follows:

$$\widetilde{M}_{sk}[\tilde{u}] = \frac{\partial u^i}{\partial \tilde{u}^s} M_{ij}[u(\tilde{u})] \frac{\partial u^j}{\partial \tilde{u}^k}. \qquad (2.10)$$

This is an elementary consequence of the transformation law of vector fields on a manifold and the following requirement of invariance for the symplectic form:

$$\omega(\xi, \eta)[\gamma] = \int_{S^1} \xi^i[u(x)] M_{ij}[u(x)] \eta^j[u(x)]\, dx =$$

$$= \int_{S^1} \tilde{\xi}^s[\tilde{u}(x)] \widetilde{M}_{sk}[\tilde{u}(x)] \tilde{\eta}^k[\tilde{u}(x)]\, dx, \qquad (2.11)$$

$$\xi^i[u(\tilde{u})] = \tilde{\xi}^s[\tilde{u}] \frac{\partial u^i}{\partial \tilde{u}^s}, \qquad \eta^j[u(\tilde{u})] = \tilde{\eta}^k[\tilde{u}] \frac{\partial u^j}{\partial \tilde{u}^k}.$$

In what follows we shall always consider symplectic operators as corresponding differential-geometric objects with respect to local changes of coordinates on a manifold M.

Note that any symplectic form $\Omega_{ij}(u)\, du^i \wedge du^j$ on a symplectic manifold M defines the following obvious non-degenerate symplectic structure on ΩM:

$$\omega(\xi, \eta)[\gamma] = \int_{S^1} \xi^i(x) \Omega_{ij}(u(x)) \eta^j(x)\, dx. \qquad (2.12)$$

Definition 2.2 Symplectic structures of the form

$$\omega(\xi, \eta)[\gamma] = \int_{S^1} \xi^i(x) \omega_{ij}(u(x)) \eta^j(x)\, dx \qquad (2.13)$$

on a loop space are called *ultralocal*.

Obviously, all ultralocal symplectic structures on the loop space ΩM of a smooth manifold M are generated by closed 2-forms $\omega_{ij}(u)$ on the manifold M. In this case the symplectic conditions (2.6) and (2.7) are equivalent to the well-known relations of classic finite-dimensional symplectic geometry

$$\omega_{ij}(u) = -\omega_{ji}(u) \quad (\text{skew} - \text{symmetry}) \qquad (2.14)$$

and

$$\frac{\partial \omega_{ij}}{\partial u^k} + \frac{\partial \omega_{jk}}{\partial u^i} + \frac{\partial \omega_{ki}}{\partial u^j} = 0 \quad (\text{closedness}). \qquad (2.15)$$

Generally speaking, a symplectic operator defining a symplectic structure on a loop space can depend not only on the functions $u^i(x)$ as in the ultralocal case but also, in particular, on the derivatives of these functions with respect to x.

Definition 2.3 A symplectic structure ω on a loop space ΩM is called *local* if it is defined by a matrix differential symplectic operator M_{ij} such that all the coefficients of this operator are functions of x, $u^i(x)$, and finitely many derivatives $u^i_{(k)}(x)$.

Let us consider an arbitrary local symplectic structure defined by a matrix differential symplectic operator

$$M_{ij} = \sum_{k=0}^{n} a_{ij}^k(x, u(x), u_x, \ldots) \frac{d^k}{dx^k}. \qquad (2.16)$$

By the *order of a local symplectic structure* we mean the order n of the corresponding differential symplectic operator M_{ij}. It is clear that the property of localization and the order of the local symplectic structure are invariant with respect to local changes of coordinates on the corresponding manifold.

Skew-symmetry of the local bilinear form (2.9), (2.16) is equivalent to the relation

$$\sum_k a_{ij}^k \frac{d^k}{dx^k} = \sum_k (-1)^{k+1} \frac{d^k}{dx^k} \circ a_{ji}^k, \qquad (2.17)$$

since for arbitrary smooth vector-functions $\xi(x)$ and $\eta(x)$ we have

$$\int_{S^1} \xi^i(x) a_{ij}^k \frac{d^k}{dx^k} (\eta^j(x)) dx = \omega(\xi, \eta) = -\omega(\eta, \xi) =$$

$$= -\int_{S^1} \eta^j(x) a^k_{ji} \frac{d^k}{dx^k}(\xi^i(x)) dx =$$

$$= -\int_{S^1} \xi^i(x)(-1)^k \frac{d^k}{dx^k}(a^k_{ji}\eta^j(x)) dx.$$

The symbol $\circ$ means operator multiplication:

$$(A \circ B)(f) = A(B(f)).$$

The closedness of the skew-symmetric local bilinear form (2.9), (2.16), (2.17) gives the following relation:

$$(d\omega)(\xi, \eta, \zeta)[\gamma] =$$

$$= \int_{S^1} \xi^i_{(s)}(x) \frac{\partial}{\partial u^i_{(s)}}\left[\eta^j(x)\left(a^r_{jk}\frac{d^r}{dx^r}(\zeta^k(x))\right)\right] dx -$$

$$- \int_{S^1} \eta^j_{(s)}(x) \frac{\partial}{\partial u^j_{(s)}}\left[\xi^i(x)\left(a^r_{ik}\frac{d^r}{dx^r}(\zeta^k(x))\right)\right] dx +$$

$$+ \int_{S^1} \zeta^k_{(s)}(x) \frac{\partial}{\partial u^k_{(s)}}\left[\xi^i(x)\left(a^r_{ij}\frac{d^r}{dx^r}(\eta^j(x))\right)\right] dx =$$

$$= \int_{S^1} \xi^i(x)(-1)^s \frac{d^s}{dx^s}\left(\eta^j(x)\frac{\partial a^r_{jk}}{\partial u^i_{(s)}}\zeta^k_{(r)}(x)\right) dx -$$

$$- \int_{S^1} \xi^i(x)\eta^j_{(s)}(x)\frac{\partial a^r_{ik}}{\partial u^j_{(s)}}\zeta^k_{(r)}(x) dx +$$

$$+ \int_{S^1} \xi^i(x)\zeta^k_{(s)}(x)\frac{\partial a^r_{ij}}{\partial u^k_{(s)}}\eta^j_{(r)}(x) dx \equiv 0,$$

that is, for arbitrary smooth vector-functions $\eta(x)$ and $\zeta(x)$ the following identity must be fulfilled:

$$(-1)^s \frac{d^s}{dx^s}\left(\frac{\partial a^r_{jk}}{\partial u^i_{(s)}}\eta^j(x)\zeta^k_{(r)}(x)\right) - \frac{\partial a^r_{ik}}{\partial u^j_{(s)}}\eta^j_{(s)}(x)\zeta^k_{(r)}(x) +$$

$$+ \frac{\partial a^r_{ij}}{\partial u^k_{(s)}}\eta^j_{(r)}(x)\zeta^k_{(s)}(x) \equiv 0. \tag{2.18}$$

2.1.3 Complete description of all local matrix symplectic structures of zero order on loop spaces of smooth manifolds

As the first non-trivial example we find the complete description of all local matrix symplectic structures of zero order on an arbitrary loop space ΩM, that is, all symplectic structures defined by an operator of multiplication by a matrix function

$$\omega_{ij}(x, u, u_x, u_{xx}, \ldots, u_{(k)}) :$$

$$\omega(\xi, \eta)[\gamma] = \int_{S^1} \xi^i \omega_{ij}(x, u, u_x, u_{xx}, \ldots, u_{(k)}) \eta^j \, dx. \qquad (2.19)$$

This is a natural generalization of the classic symplectic structures $\omega_{ij}(u)$ on a symplectic manifold.

Remark 2.2 In what follows we shall usually omit the symbol of the loop γ in the notation of a symplectic structure $\omega(\xi, \eta)$ considered at an arbitrary point (a loop) $\gamma = \{u^i(x),\ 1 \leq i \leq N,\ x \in S^1\} \in \Omega M$.

Lemma 2.1 *The operator* $\omega_{ij}(x, u, u_x, u_{xx}, \ldots, u_{(k)})$ *is symplectic if and only if*

1. $\omega_{ij} = -\omega_{ji}$ *(skew-symmetry)*;

2. *for two arbitrary smooth vector-functions*

$$\eta(x) = (\eta^1(x), \ldots, \eta^N(x)) \quad \text{and} \quad \zeta(x) = (\zeta^1(x), \ldots, \zeta^N(x))$$

the following identity (the condition of closedness) is fulfilled:

$$(-1)^n \left(\frac{d}{dx} \right)^n \left(\frac{\partial \omega_{ij}}{\partial u^s_{(n)}} \eta^i(x) \zeta^j(x) \right) + \frac{\partial \omega_{si}}{\partial u^j_{(n)}} \eta^i(x) \zeta^j_{(n)}(x) -$$

$$- \frac{\partial \omega_{sj}}{\partial u^i_{(n)}} \eta^i_{(n)}(x) \zeta^j(x) \equiv 0. \qquad (2.20)$$

Lemma 2.1 follows from the general relations (2.17) and (2.18).

Proposition 2.1 ([124], [115]) *The operator*

$$\omega_{ij}(u, u_x, u_{xx}, \ldots, u_{(k)}),$$

which does not depend explicitly on the independent variable x (that is, the operator is translation invariant), defines a symplectic (presymplectic) form (2.19) on ΩM if and only if

$$\omega_{ij}(u, u_x, u_{xx}, \ldots, u_{(k)}) = T_{ijk}(u)u_x^k + \Omega_{ij}(u), \qquad (2.21)$$

where $T_{ijk}(u)$ is an arbitrary closed 3-form on the manifold M and $\Omega_{ij}(u)$ is an arbitrary closed 2-form on M.

Proposition 2.2 ([124], [115]) *In the most general case, an arbitrary local matrix operator of zero order*

$$\omega_{ij}(x, u, u_x, u_{xx}, \ldots, u_{(k)})$$

defines a symplectic (presymplectic) form (2.19) on ΩM if and only if

$$\omega_{ij}(x, u, u_x, u_{xx}, \ldots, u_{(k)}) =$$
$$= \left(\int_0^x (d\Omega)_{ijk}(z, u(x))\, dz + S_{ijk}(u(x)) \right) u_x^k +$$
$$+\Omega_{ij}(x, u(x)), \qquad (2.22)$$

where $S_{ijk}(u)$ is an arbitrary closed 3-form on the manifold M and $\Omega(z, u)$ is an arbitrary one-parameter family of 2-forms on M. The differential d in (2.22) is defined by the standard formula

$$(d\Omega)_{ijk}(z, u(x)) = \frac{\partial \Omega_{ij}(z, u(x))}{\partial u^k} +$$
$$+\frac{\partial \Omega_{ki}(z, u(x))}{\partial u^j} + \frac{\partial \Omega_{jk}(z, u(x))}{\partial u^i} . \qquad (2.23)$$

Proof. Assume that there exist values of the indices i, j, s and $m \geq 2$ such that $\partial \omega_{ij}/\partial u_{(m)}^s \neq 0$.
Let

$$R = \max\{r : \text{ there exist } p, q, l \text{ such that } \partial \omega_{pq}/\partial u_{(r)}^l \neq 0\}.$$

By our assumption we have $R \geq 2$. The coefficient of $\eta_x^i \zeta_{(R-1)}^j$ in the identity (2.20) is equal to $(-1)^R R \, \partial \omega_{ij}/\partial u_{(R)}^s$ and by virtue of the identity (2.20) this coefficient must be equal to zero, that is, $\partial \omega_{ij}/\partial u_{(R)}^s \equiv 0$ for all i, j, s. Thus it is proved that $R \leq 1$ and therefore $\omega_{ij} = \omega_{ij}(x, u, u_x)$ are functions depending only on x, $u^i(x)$, and the first-order derivatives $u_x^i(x)$. In this case, the identity (2.20) can be rewritten in the following way:

$$\frac{\partial \omega_{ij}}{\partial u^s} \eta^i \zeta^j - \left(\frac{\partial \omega_{ij}}{\partial u_x^s} \eta^i \zeta^j \right)_x + \frac{\partial \omega_{si}}{\partial u^j} \eta^i \zeta^j +$$
$$+ \frac{\partial \omega_{si}}{\partial u_x^j} \eta^i \zeta_x^j - \frac{\partial \omega_{sj}}{\partial u^i} \eta^i \zeta^j - \frac{\partial \omega_{sj}}{\partial u_x^i} \eta_x^i \zeta^j \equiv 0,$$

whence, by considering the coefficient of $u_{xx}^r(x) \eta^i \zeta^j$, we immediately deduce that

$$\frac{\partial^2 \omega_{ij}}{\partial u_x^s \partial u_x^r} \equiv 0,$$

that is,

$$\omega_{ij} = T_{ijk}(x, u(x)) u_x^k + \Omega_{ij}(x, u(x)). \tag{2.24}$$

Thus, local matrix symplectic operators of zero order, that is, all symplectic operators of the form $\omega_{ij}(x, u, u_x, u_{xx}, \ldots, u_{(k)})$, depend only on x, $u^i(x)$, $u_x^i(x)$ and must be quasilinear with respect to the first-order derivatives $u_x^i(x)$.

For operators ω_{ij} of the form (2.24) the skew-symmetry and the identity (2.20) are equivalent to the following relations for the coefficients $T_{ijk}(x, u(x))$ and $\Omega_{ij}(x, u(x))$:

$$T_{ijk} = -T_{jik}, \tag{2.25}$$

$$\Omega_{ij} = -\Omega_{ji}, \tag{2.26}$$

$$T_{ijk} = T_{kij}, \tag{2.27}$$

$$\frac{\partial T_{ijk}}{\partial u^s} - \frac{\partial T_{ijs}}{\partial u^k} + \frac{\partial T_{jsk}}{\partial u^i} - \frac{\partial T_{isk}}{\partial u^j} = 0, \tag{2.28}$$

$$\frac{\partial T_{ijs}}{\partial x} = \frac{\partial \Omega_{ij}}{\partial u^s} + \frac{\partial \Omega_{si}}{\partial u^j} + \frac{\partial \Omega_{js}}{\partial u^i}. \tag{2.29}$$

Relations (2.25)–(2.29) are equivalent to the following differential-geometric conditions: $\Omega_{ij}(x,u)$ is a one-parameter family of skew-symmetric 2-forms on the manifold M, which is parametrized by the circle S^1 (here $x \in S^1$ is a formal parameter),

$$T_{ijk}(x,u) = \int_0^x (d\Omega)_{ijk}(z,u)\,dz + S_{ijk}(u), \qquad (2.30)$$

where $S_{ijk}(u)$ is an arbitrary closed 3-form on the manifold M (it is skew-symmetric with respect to any pair of indices),

$$(d\Omega)_{ijk}(z,u) = \frac{\partial \Omega_{ij}(z,u)}{\partial u^k} + \frac{\partial \Omega_{ki}(z,u)}{\partial u^j} + \frac{\partial \Omega_{jk}(z,u)}{\partial u^i}, \qquad (2.31)$$

that is, $T_{ijk}(x,u)$ is a one-parameter family of closed 3-forms on the manifold M, which is defined by formula (2.30).

These conditions have a very simple differential-geometric sense in the translation invariant case, that is, if the functions ω_{ij} do not depend explicitly on the independent variable x. In this case,

$$\omega_{ij}(u, u_x, u_{xx}, \ldots, u_{(k)}) = T_{ijk}(u)u_x^k + \Omega_{ij}(u), \qquad (2.32)$$

where $T_{ijk}(u)$ is an arbitrary closed 3-form on the manifold M and $\Omega_{ij}(u)$ is an arbitrary closed 2-form on M.

Note that in Propositions 2.1 and 2.2 non-degeneracy of the operator ω_{ij} is not assumed.

The following statement is an elementary consequence of Proposition 2.1: *on the loop space of any two-dimensional manifold any local translation-invariant symplectic structure of zero order is ultralocal.*

Corollary 2.1 ([115]) *All local symplectic (presymplectic) forms of zero order on ΩM, that is, forms written as (2.19), that are invariant with respect to the group $\mathrm{Diff}^+(S^1)$ of orientation-preserving diffeomorphisms of the circle S^1 have the form*

$$\omega(\xi, \eta) = \int_{S^1} \xi^i T_{ijk}(u(x))u_x^k \eta^j\,dx, \qquad (2.33)$$

where $T_{ijk}(u)$ is an arbitrary closed 3-form on the manifold M, and they are always degenerate closed 2-forms on ΩM, since

$$\det(T_{ijk}(u)u_x^k) \equiv 0.$$

In particular, the velocity vector field $\nu = \{u_x^i\}$ of a loop $\gamma = \{u^i(x)\}$ always lies in the null space of the 2-form (2.33).

In fact,

$$T_{ijk} u_x^k u_x^j = -T_{ikj} u_x^k u_x^j = -T_{ijk} u_x^j u_x^k,$$

that is $T_{ijk} u_x^k u_x^j \equiv 0$, and consequently $\{u_x^j\} \in \mathrm{Ker}(T_{ijk} u_x^k)$.

In the theory of integrable systems a very important role is played by invertible symplectic operators, since the corresponding inverse operators generate Poisson structures (in this case, they are not necessarily defined on the whole loop space ΩM). In particular, if a closed 2-form $\Omega_{ij}(u)$ is non-degenerate (that is, it is a symplectic form on the manifold M), then for any closed 3-form $T_{ijk}(u)$ on the symplectic manifold M we always have

$$\det(T_{ijk}(u)u_x^k + \Omega_{ij}(u)) \not\equiv 0,$$

and consequently there formally exists an inverse operator defined, generally speaking, not on the whole loop space ΩM.

Now let $\Omega_{ij}(u)$ be an arbitrary 2-form on a manifold M. Then formula (2.22) defines the following multivalued symplectic operator of zero order on ΩM:

$$\omega_{ij}(x, u, u_x) = [(x + 2m\pi)(d\Omega)_{ijk}(u(x)) +$$
$$+ S_{ijk}(u(x))]u_x^k + \Omega_{ij}(u(x)), \quad m \in \mathbf{Z}, \qquad (2.34)$$

where $S_{ijk}(u)$ is an arbitrary closed 3-form on the manifold M.

If $\Omega_{ij}(u)$ is a non-degenerate 2-form $(\det \Omega_{ij}(u) \neq 0)$, that is, the manifold (M, Ω_{ij}) is *almost symplectic*, then the multivalued symplectic structure (2.34) is formally invertible

$$(\det \omega_{ij}(x, u, u_x) \not\equiv 0)$$

for any closed 3-form $S_{ijk}(u)$ (in particular, for the zero form) on the manifold M.

It is also possible to construct single-valued formally invertible symplectic structures on the loop space of an arbitrary almost symplectic manifold. An arbitrary smooth (possibly, multivalued) function on the circle $f(x)$, $x \in S^1$, which is not constant on arbitrary intervals of this circle, and any closed 3-form $S_{ijk}(u)$ on an almost symplectic manifold (M, Ω_{ij}) generate the following formally invertible symplectic structure on the loop space ΩM:

$$\omega_{ij}(x, u, u_x) = [f(x)(d\Omega)_{ijk}(u(x)) +$$
$$+ S_{ijk}(u(x))]u_x^k + f'(x)\Omega_{ij}(u(x)). \qquad (2.35)$$

In particular,

$$\omega_{ij}(x, u, u_x) = \sin x \, (d\Omega)_{ijk}(u(x))u_x^k + \cos x \, \Omega_{ij}(u(x))$$

is a simple example of a formally invertible symplectic structure on the loop space of an arbitrary almost symplectic manifold (M, Ω_{ij}).

The formally invertible symplectic structures constructed above on loop spaces of almost symplectic manifolds are not translation-invariant.

2.1.4   Poisson structures

For an arbitrary functional I on a loop space ΩM, the variational derivative of this functional is defined just as in the classic calculus of variations by the equality

$$\delta I \equiv I[u + \delta u] - I[u] = \int_{S^1} \frac{\delta I}{\delta u^k(x)} \delta u^k(x) \, dx + o(\delta u). \qquad (2.36)$$

In particular, for *local functionals*, that is, for functionals of the form

$$I = \int_{S^1} L(x, u, u_x, u_{xx}, \ldots, u_{(n)}) \, dx, \qquad (2.37)$$

the variational derivative has the form

$$\frac{\delta I}{\delta u^k(x)} = E(L) \equiv \frac{\partial L}{\partial u^k} - \frac{d}{dx} \frac{\partial L}{\partial u_x^k} +$$
$$+ \frac{d^2}{dx^2} \frac{\partial L}{\partial u_{xx}^k} - \cdots + (-1)^r \frac{d^r}{dx^r} \frac{\partial L}{\partial u_{(r)}^k} + \cdots. \qquad (2.38)$$

The linear operator

$$E = \frac{\partial}{\partial u^k} - \frac{d}{dx} \frac{\partial}{\partial u_x^k} + \frac{d^2}{dx^2} \frac{\partial}{\partial u_{xx}^k} - \cdots + (-1)^r \frac{d^r}{dx^r} \frac{\partial}{\partial u_{(r)}^k} + \cdots$$

is called the *Euler operator*.

Definition 2.4 A linear operator $K = (K^{ij})$ depending on loops $\gamma \in \Omega M$:

$$K[\gamma]: \quad T_\gamma^* \Omega M \to T_\gamma \Omega M, \qquad (2.39)$$

is called a *Hamiltonian* operator or *Hamiltonian* (*Poisson*) structure on the loop space ΩM if it defines on ΩM the following Poisson bracket:

$$\{u^i(x), u^j(y)\} = K^{ij}[u(x)]\delta(x - y), \qquad (2.40)$$

that is, for arbitrary functionals F, G, H on the loop space ΩM

$$\{F, G\} = \int_{S^1} \frac{\delta F}{\delta u^i(x)} K^{ij}[u(x)] \frac{\delta G}{\delta u^j(x)} \, dx \qquad (2.41)$$

so that the bracket (2.41) is skew-symmetric:

$$\{F, G\} = -\{G, F\}, \qquad (2.42)$$

and satisfies the Jacobi identity

$$\{F, \{G, H\}\} + \{G, \{H, F\}\} + \{H, \{F, G\}\} = 0. \qquad (2.43)$$

For the bracket (2.41) the following Leibniz identity is automatically fulfilled:

$$\begin{aligned}
\{f(u(x))g(u(y)), h(u(z))\} &= \\
= f(u(x))\{g(u(y)), h(u(z))\} &+ \\
+ g(u(y))\{f(u(x)), h(u(z))\} & \qquad (2.44)
\end{aligned}$$

or

$$\{FG, H\} = F\{G, H\} + G\{F, H\}. \qquad (2.45)$$

We shall often use the field-theoretic representation (2.40) for the Poisson bracket (2.41) defined on point functionals $F[u(z)] = u^i(x)$, $G[u(z)] = u^j(y)$. If we define a Poisson bracket only on point functionals by formula (2.40), then it is necessary to require the fulfilment of skew-symmetry, the Jacobi identity and the Leibniz identity on arbitrary point functionals.

In order to guarantee that a skew-symmetric operator is Hamiltonian it is sufficient to require the fulfilment of the Jacobi identity only on all linear functionals of the form

$$I = \int_{S^1} f_i(x) u^i(x)\, dx, \tag{2.46}$$

where $f_i(x)$ are arbitrary functions (see [101]–[103]). In [102] it is shown by the present author that this condition is equivalent to the well-known Hamiltonian criterion for an arbitrary skew-symmetric operator, namely, to the condition that its Schouten bracket is equal to zero (see Gelfand, Dorfman [68]).

Under a local change of coordinates $u^i = u^i(\tilde{u})$ on a manifold M any Hamiltonian operator $K^{ij}[u(x)]$ is transformed as follows:

$$\widetilde{K}^{sk}[\tilde{u}] = \frac{\partial \tilde{u}^s}{\partial u^i} K^{ij}[u(\tilde{u})] \frac{\partial \tilde{u}^k}{\partial u^j}.$$

This is an elementary consequence of the transformation law for variational derivatives and the requirement of invariance of the corresponding Poisson bracket:

$$\{F, G\} = \int_{S^1} \frac{\delta F}{\delta u^i(x)} K^{ij}[u(x)] \frac{\delta G}{\delta u^j(x)}\, dx =$$

$$= \int_{S^1} \frac{\delta F}{\delta \tilde{u}^s(x)} \widetilde{K}^{sk}[\tilde{u}(x)] \frac{\delta G}{\delta \tilde{u}^k(x)}\, dx,$$

$$\frac{\delta F}{\delta \tilde{u}^s(x)} = \frac{\delta F}{\delta u^i(x)} \frac{\partial u^i}{\partial \tilde{u}^s},$$

$$\frac{\delta G}{\delta \tilde{u}^s(x)} = \frac{\delta G}{\delta u^i(x)} \frac{\partial u^i}{\partial \tilde{u}^s}.$$

In what follows we shall always consider Hamiltonian operators as corresponding differential-geometric objects with respect to local changes of coordinates on a manifold M.

Definition 2.5 A Poisson structure is called *local* if the Hamiltonian operator which defines this Poisson structure is a differential operator with coefficients depending only on the independent variable x, the field variables $u^i(x)$, and finitely many derivatives

$u^i_{(k)}(x)$, and it is called *ultralocal* if the Hamiltonian operator has the form

$$K^{ij} = \omega^{ij}(u(x)).$$

By the *order of a local Poisson structure* we mean the order of the corresponding Hamiltonian differential operator.

The property of localization and the order of a local Poisson structure are invariant with respect to local changes of coordinates on the corresponding manifold.

Ultralocal Poisson structures on a loop space ΩM are exactly generated by classic Poisson structures on the finite-dimensional manifold M.

In the finite-dimensional case, if $\det(\omega_{ij}) \neq 0$ everywhere on a manifold M, then $\omega_{ij}(u)$ is a symplectic structure on M if and only if the inverse matrix $\omega^{ij}(u)$, $\omega^{is}(u)\omega_{sj}(u) = \delta^i_j$, defines a Poisson structure $\omega^{ij}(u)$ on the manifold M:

$$\{u^i, u^j\} = \omega^{ij}(u). \qquad (2.47)$$

We recall that a tensor $\omega^{ij}(u)$ on a manifold M defines a *Poisson structure* (or a *Poisson bivector*) if the following bilinear operation (a *Poisson bracket*) is given on the space of all smooth functions on the manifold:

$$\{f(u), g(u)\} = \omega^{ij}(u)\frac{\partial f}{\partial u^i}\frac{\partial g}{\partial u^j}, \qquad (2.48)$$

such that this bracket is skew-symmetric

$$\{f, g\} = -\{g, f\} \qquad (2.49)$$

and satisfies the Jacobi identity

$$\{\{f, g\}, h\} + \{\{g, h\}, f\} + \{\{h, f\}, g\} = 0 \qquad (2.50)$$

for arbitrary smooth functions $f(u), g(u), h(u)$ on the manifold.

On M any Poisson bivector $\omega^{ij}(u)$ is defined by the relations

$$\omega^{ij} = -\omega^{ji} \quad \text{(skew} - \text{symmetry)}, \qquad (2.51)$$

$$\frac{\partial \omega^{ij}}{\partial u^s}\omega^{sk} + \frac{\partial \omega^{jk}}{\partial u^s}\omega^{si} + \frac{\partial \omega^{ki}}{\partial u^s}\omega^{sj} = 0 \quad \text{(the Jacobi identity)}. \qquad (2.52)$$

In the infinite-dimensional case, if an operator M_{ij} is invertible, then in just the same way as in the finite-dimensional case the following general statement is valid: M_{ij} is a symplectic operator if and only if the operator K^{ij} which is the inverse operator to M_{ij}, that is, $K^{ij}M_{jk} = \delta^i_k$, is Hamiltonian, that is, it defines a Poisson bracket (a Poisson structure) of the form (2.41).

Let an operator M_{ij} be invertible, $K^{ij}M_{jk} = \delta^i_k$. Consider vector fields ξ, η, and ζ of the special form

$$\xi^i = K^{is}f_s(x), \quad \eta^j = K^{js}g_s(x), \quad \zeta^r = K^{rs}h_s(x),$$

where $f_s(x), g_s(x)$, and $h_s(x), 1 \leq s \leq N$, are arbitrary functions. Then the condition that for a skew-symmetric bilinear form (2.9) the relation $(d\omega)(\xi, \eta, \zeta) = 0$ on vector fields of the special form is equivalent to the Jacobi identity on arbitrary linear functionals

$$I_f = \int_{S^1} f_i(x)u^i(x)\,dx, \quad I_g = \int_{S^1} g_i(x)u^i(x)\,dx,$$

$$\text{and} \quad I_h = \int_{S^1} h_i(x)u^i(x)\,dx$$

for the bracket

$$\{F, G\}_K = \int_{S^1} \frac{\delta F}{\delta u^i(x)} K^{ij}[u(x)] \frac{\delta G}{\delta u^j(x)}\,dx$$

defined by the skew-symmetric operator K^{ij}:

$$(d\omega)(\xi, \eta, \zeta) = \sum_{(\xi,\eta,\zeta)} \left\{ \int_{S^1} \xi^i \frac{\delta\omega(\eta,\zeta)}{\delta u^i(x)}\,dx - \omega([\xi,\eta],\zeta) \right\} =$$

$$= \sum_{(f,g,h)} \left\{ \int_{S^1} (K^{is}f_s) \frac{\delta \int_{S^1}(K^{jl}g_l)M_{jr}(K^{rp}h_p)\,dx}{\delta u^i(x)}\,dx \right\} -$$

$$- \sum_{(\xi,\eta,\zeta)} \omega(\eta'_u\xi - \xi'_u\eta, \zeta) =$$

$$= \sum_{(f,g,h)} \left\{ \int_{S^1} (K^{is}f_s) \frac{\delta \int_{S^1}(K^{jl}g_l)h_j\,dx}{\delta u^i(x)}\,dx - \right.$$

$$- \frac{d}{d\varepsilon} \left[\int_{S^1} (K^{jl}[u^i + \varepsilon(K^{is}f_s)]g_l)M_{jr}(K^{rp}h_p)\,dx - \right.$$

$$-\int_{S^1}(K^{is}[u^j+\varepsilon(K^{jl}g_l)]f_s)M_{ir}(K^{rp}h_p)\,dx\Big]\Big|_{\varepsilon=0}\Big\}=$$

$$=\sum_{(f,g,h)}\Big\{\int_{S^1}(K^{is}f_s)\frac{\delta\{I_h,I_g\}_K}{\delta u^i(x)}\,dx-$$

$$-\frac{d}{d\varepsilon}\Big[\int_{S^1}h_j(K^{jl}[u^i+\varepsilon(K^{is}f_s)]g_l)\,dx-$$

$$-\int_{S^1}h_i(K^{is}[u^j+\varepsilon(K^{jl}g_l)]f_s)\,dx\Big]\Big|_{\varepsilon=0}\Big\}=$$

$$=\sum_{(f,g,h)}\Big\{\{\{I_h,I_g\}_K,I_f\}_K-$$

$$-\Big[\int_{S^1}\frac{\delta\int_{S^1}h_j(K^{jl}g_l)\,dx}{\delta u^i(x)}(K^{is}f_s)\,dx-$$

$$-\int_{S^1}\frac{\delta\int_{S^1}h_j(K^{js}f_s)\,dx}{\delta u^i(x)}(K^{il}g_l)\,dx\Big]\Big\}=$$

$$=\sum_{(f,g,h)}\{\{I_h,I_f\}_K,I_g\}_K.$$

Thus, under the additional condition of non-degeneracy of the operator

$$\omega_{ij}(x,u,u_x,u_{xx},\ldots,u_{(k)}):\quad \det(\omega_{ij})\neq 0,$$

Proposition 2.2 also gives a complete description of all non-degenerate local matrix Hamiltonian operators of zero order, that is, Hamiltonian operators of the form

$$K^{ij}=\omega^{ij}(x,u,u_x,u_{xx},\ldots,u_{(k)}),\quad \det(\omega^{ij})\neq 0.\quad(2.53)$$

Moreover, from Propositions 2.1 and 2.2 we immediately obtain the following important corollaries.

Corollary 2.2 ([124], [115]) *Any closed 3-form*

$$T_{ijk}(u)\,du^i\wedge du^j\wedge du^k$$

on a symplectic manifold (M,Ω_{ij}) *defines the Poisson structure*

$$\{F,G\}=\int_{S^1}\frac{\delta F}{\delta u^i(x)}\omega^{ij}(u,u_x)\frac{\delta G}{\delta u^j(x)}\,dx,$$

$$\omega^{si}(u,u_x)(T_{ijk}(u)u_x^k+\Omega_{ij}(u))=\delta_j^s,\quad(2.54)$$

where F and G are arbitrary functionals on ΩM.

Proposition 2.1 describes all non-degenerate translation-invariant local matrix Poisson structures of zero order (in the non-degenerate case $\det(\omega_{ij}) \neq 0$).

Corollary 2.3 ([115]) *Let $T_{ijk}(u) = \text{const}$ and $\Omega_{ij} = \text{const}$ be arbitrary tensors that are skew-symmetric with respect to any pair of indices and constant in a fixed local coordinate system $u^1, \ldots, u^N$ on a manifold M. Then in formula (2.54) we have $\omega^{si} = \omega^{si}(u_x)$, and after the transformation $v^i = u^i_x$, $i = 1, \ldots, N$, the Poisson bracket (2.54) reduces to the form*

$$\{v^i(x), v^j(y)\} = -\frac{d}{dx}\omega^{ij}(v(x))\frac{d}{dx}\delta(x-y),$$
$$\omega^{si}(v(x))(T_{ijk}v^k(x) + \Omega_{ij}) = \delta^s_j, \qquad (2.55)$$

that is, it becomes a local homogeneous Poisson bracket of the second order.

As proved in [172], [26], any homogeneous Poisson structure of the second order can be reduced to the bracket (2.55) by a local change of coordinates on the manifold (see Section 4.5.2). Thus, Corollary 2.3 shows that all local homogeneous Poisson structures of the second order are generated by special local symplectic structures of zero order. This curious fact was not noticed in the works [172], [173], [26] on classifications of homogeneous Poisson structures of the second order.

Remark 2.3 If $v^i(x) = u^i_x(x)$, then for any functional H the following relation is valid:

$$\frac{\delta H}{\delta u^i(x)} = -\frac{d}{dx}\frac{\delta H}{\delta v^i(x)}. \qquad (2.56)$$

Corollary 2.4 *Any closed 3-form $S_{ijk}(u)\, du^i \wedge du^j \wedge du^k$ (in particular, the zero form) on almost symplectic manifold (M, Ω_{ij}) defines non-trivial Poisson structures generated by the invertible symplectic structures (2.34) and (2.35).*

The problem of classifying non-degenerate Hamiltonian operators of the form (2.53) was also considered in [186], [6]. In [6] a local classification of such Hamiltonian operators with respect to Lie transformations was obtained, but their explicit description and general form were not studied (in [186] it was claimed by mistake that Hamiltonian operators of the form (2.53) cannot depend on derivatives of the field variables $u^i(x)$).

The problem of a differential-geometric description of degenerate Poisson brackets given by local matrix Hamiltonian operators of zero order, that is, of the form

$$K^{ij} = \omega^{ij}(x, u, u_x, u_{xx}, \ldots, u_{(k)}), \quad \det(\omega^{ij}) = 0, \qquad (2.57)$$

is more complicated and it is not yet solved. It is easy to show that a local Hamiltonian operator of the form (2.57) can depend, in contrast to symplectic operators of the same form (see Propositions 2.1 and 2.2), on derivatives of any orders, but the description of such operators, that is, at least the structure of possible dependence on derivatives of the field variables $u^i(x)$, has not yet been obtained even in the simplest case if $N = 3$ and the operator has the form

$$(\omega^{ij}(x, u, u_x, u_{xx}, \ldots)) = \begin{pmatrix} 0 & \omega^1 & \omega^2 \\ -\omega^1 & 0 & \omega^3 \\ -\omega^2 & -\omega^3 & 0 \end{pmatrix}.$$

Definition 2.6 A system of equations is called *Hamiltonian* if it can be written in the form

$$u_t^i = K^{ij}[u(x)]\frac{\delta H}{\delta u^j(x)} \equiv \{u^i(x), H\}, \qquad (2.58)$$

where $K^{ij}[u(x)]$ is a Hamiltonian operator, $\{\cdot, \cdot\}$ is the corresponding Poisson bracket defined by the operator K^{ij}, and H is a functional (the *Hamiltonian* of the system). Correspondingly, a system of equations is called *symplectic* if it can be written in the form

$$M_{ij}[u(x)]u_t^j = \frac{\delta H}{\delta u^i(x)}, \qquad (2.59)$$

where $M_{ij}[u(x)]$ is a symplectic operator and H is the Hamiltonian, or, equivalently, in the form

$$\omega(\delta u, u_t) = \delta H, \qquad (2.60)$$

where ω is the corresponding symplectic structure on a loop space, and relation (2.60) is fulfilled for arbitrary variations $\delta u^i(x)$ of a loop $\gamma = \{u^i(x)\}$ on a manifold M, and H is a functional on the loop space ΩM.

2.1.5 Lagrangian description of local symplectic structures of zero order

Let us return to the symplectic structures (2.21). Locally, in each coordinate neighbourhood on a manifold any matrix local translation invariant symplectic structures of zero order can be represented in the following way (here we apply the classic Poincaré lemma for closed differential forms on a manifold to formula (2.21)):

$$\omega_{ij} = \left(\frac{\partial a_{ji}}{\partial u^k} + \frac{\partial a_{kj}}{\partial u^i} + \frac{\partial a_{ik}}{\partial u^j} \right) u_x^k + \frac{\partial b_j}{\partial u^i} - \frac{\partial b_i}{\partial u^j} \,. \qquad (2.61)$$

It is interesting that all symplectic structures of the form (2.61) generate Hamiltonian (or symplectic) representations for motion equations of non-linear sigma-models with purely skew-symmetric metrics.

Proposition 2.3 ([124], [115]) *Lagrangian systems*

$$\frac{\delta S}{\delta u^i(x)} = 0$$

given by actions of the form

$$S = \int \left(a_{ij}(u) u_x^i u_t^j + b_i(u) u_t^i + \right.$$
$$\left. + U(x, t, u, u_x, u_{xx}, \ldots, u_{(k)}) \right) dx\, dt, \qquad (2.62)$$

where $a_{ij}(u)$ is an arbitrary covariant bivector on a manifold, that is, it is an arbitrary skew-symmetric covariant tensor $a_{ij} = -a_{ji}$, $b_i(u)$ is an arbitrary covector, and $U(x, t, u, u_x, u_{xx}, \ldots, u_{(k)})$ is an arbitrary function, are symplectic (Hamiltonian) with respect to symplectic structures of the form (2.61). The converse statement is also true, namely, any Hamiltonian system with a symplectic structure of the form (2.61) and with Hamiltonian $H = -\int U\, dx$ is a Lagrangian system given by an action of the form (2.62).

In fact, this statement follows from the explicit expression for the variational derivative of the action (2.62):

$$
\frac{\delta S}{\delta u^i(x)} = \omega_{ij} u_t^j - \frac{\delta H}{\delta u^i(x)} =
$$

$$
= \left[\left(\frac{\partial a_{ji}}{\partial u^k} + \frac{\partial a_{kj}}{\partial u^i} + \frac{\partial a_{ik}}{\partial u^j} \right) u_x^k + \right.
$$

$$
+ \left. \frac{\partial b_j}{\partial u^i} - \frac{\partial b_i}{\partial u^j} \right] u_t^j - \frac{\delta H}{\delta u^i(x)}, \tag{2.63}
$$

where $H = - \int U \, dx$ is the Hamiltonian.

Thus, all symplectic structures of the form (2.21) locally have a Lagrangian origin. General local symplectic structures of zero order (2.22) are also generated by actions of the form (2.62), but they depend explicitly on the independent variable x (note that this is true only locally).

In Section 2.2 we shall consider symplectic structures connected with motion equations of general two-dimensional non-linear sigma-models and present examples of bi-Hamiltonian representations of differential-geometric type for some non-linear sigma-models.

2.1.6   Compatible Poisson and symplectic structures

Definition 2.7 Two Hamiltonian operators M and K are called *compatible* if any linear combination of them, $\lambda M + \mu K$, where λ and μ are arbitrary constants, is also a Hamiltonian operator. Correspondingly, two Poisson structures (Poisson brackets) are compatible if they are given by compatible Hamiltonian operators. Invertible symplectic operators and the corresponding symplectic structures are called *compatible* if the Hamiltonian operators generated by these invertible symplectic operators are compatible.

Compatible Poisson structures were considered for the first time by Magri [88], who also explained their important role in the theory of integrable systems.

For any skew-symmetric contravariant tensor ω^{ij} on a manifold M the left-hand side of the relation (2.52) is a trivalent tensor with upper indices, which is called the *Schouten bracket of the contravariant bivector* ω^{ij}:

$$[\omega, \omega]^{ijk} = \frac{\partial \omega^{ij}}{\partial u^s} \omega^{sk} + \frac{\partial \omega^{jk}}{\partial u^s} \omega^{si} + \frac{\partial \omega^{ki}}{\partial u^s} \omega^{sj}.$$

The *Schouten bracket of two contravariant bivectors* ω^{ij} and f^{ij} is by definition the trivalent contravariant tensor

$$[\omega, f]^{ijk} = \sum_{(i,j,k)} \left(\frac{\partial \omega^{ij}}{\partial u^s} f^{sk} + \frac{\partial f^{ij}}{\partial u^s} \omega^{sk} \right). \qquad (2.64)$$

The finite-dimensional Poisson structures ω^{ij} and f^{ij} are compatible if and only if the Schouten bracket (2.64) is equal to zero.

The generalization of the Schouten bracket to the infinite-dimensional case of arbitrary skew-symmetric differential operators M^{ij} and K^{ij}, and also the proof of the corresponding criteria: a skew-symmetric differential operator is Hamiltonian if and only if its Schouten bracket is equal to zero and a pair of Hamiltonian differential operators is compatible if and only if the Schouten bracket of these operators is equal to zero, can be found in [68], [69], [20].

A system of equations is called *bi-Hamiltonian* if it can be represented in Hamiltonian form (2.58) by means of two different and linearly independent, but compatible, Hamiltonian operators $K = (K^{ij})$ and $L = (L^{ij})$:

$$u_t^i = K^{ij}[u(x)] \frac{\delta H_1}{\delta u^j(x)} = L^{ij}[u(x)] \frac{\delta H_2}{\delta u^j(x)}, \qquad (2.65)$$

where H_1 and H_2 are corresponding Hamiltonians.

Let us mention briefly here the Magri scheme associated with bi-Hamiltonian systems (see also [88]–[90], [67], [69], [20], [60], [168]).

If at least one of the Hamiltonian operators K and L is invertible (to be definite, suppose that the operator L is invertible), then the operator $R = (R_s^i) = (K^{ij}(L^{-1})_{js})$ is a recursion operator for the system (2.65), that is, it takes any symmetry of the system to a symmetry again, and, moreover, this recursion operator R generates a hierarchy of commuting Hamiltonian systems $(t = t_0)$

$$u_{t_n}^i = (R^n L)^{ij} \frac{\delta H_2}{\delta u^j(x)}, \qquad (2.66)$$

where the operator $R^n L$ is a Hamiltonian operator for any n. Furthermore, there is the following recursion relation for the conservation laws (the integrals of motion) H_m of the system (2.65):

$$L^{ij}[u(x)]\frac{\delta H_m}{\delta u^j(x)} = K^{ij}[u(x)]\frac{\delta H_{m-1}}{\delta u^j(x)}. \qquad (2.67)$$

For any m the functional H_m is a conservation law (an integral of motion) for any system from the hierarchy (2.66). All these integrals of motion H_m are in involution with respect to Poisson structures generated by the Hamiltonian operators L and K:

$$\int \frac{\delta H_m}{\delta u^i(x)} L^{ij}[u(x)]\frac{\delta H_r}{\delta u^j(x)}\, dx =$$
$$= \int \frac{\delta H_m}{\delta u^i(x)} K^{ij}[u(x)]\frac{\delta H_r}{\delta u^j(x)}\, dx = 0. \qquad (2.68)$$

Any system from the hierarchy (2.66) is bi-Hamiltonian and it can be represented as follows:

$$u^i_{t_n} = L^{ij}\frac{\delta H_{n+2}}{\delta u^j(x)} = K^{ij}\frac{\delta H_{n+1}}{\delta u^j(x)}. \qquad (2.69)$$

Actually any bi-Hamiltonian system is n-Hamiltonian for any n, since all the equations of the hierarchy (2.66) are Hamiltonian with respect to the Hamiltonian operator $R^n L$ for any n, and all these Hamiltonian operators are compatible among themselves, namely, any linear combination of the operators $R^n L$ for different n is always a Hamiltonian operator.

Generally speaking, the integrals of motion H_m obtained by the recursion relation (2.67) can be linearly dependent, but, in many cases, the recursion operator changes the order of the system, and thus the order of the highest derivative of the density of the Hamiltonian is changed (after a maximal lowering of the order by adding any total derivatives with respect to the independent variable x to the density of the Hamiltonian), which usually makes it possible to establish their independence for many concrete examples.

The recursion procedure (2.67) is usually also called the Lenard–Magri scheme, since, for the Korteweg–de Vries equation, a recursion relation of the type (2.67) for the Kruskal integrals of motion of the Korteweg–de Vries equation was found for the first time apparently by Lenard but without any connection with the theory of Hamiltonian operators (see [64], [86]).

All the statements of the Magri scheme are proved in the framework of the formal calculus of variations under the assumption that at each step the recursion relation

$$L\xi_m = K\xi_{m-1},$$

where

$$(\xi_1)_j = \frac{\delta H_1}{\delta u^j(x)}, \qquad (\xi_2)_j = \frac{\delta H_2}{\delta u^j(x)},$$

can be solved for ξ_m.

2.2 Homogeneous symplectic structures of the first order on loop spaces of pseudo-Riemannian manifolds and two-dimensional non-linear sigma-models with torsion

2.2.1 Homogeneous symplectic forms of the first order

Let us consider symplectic operators given by arbitrary non-degenerate homogeneous ordinary differential operators of the first order, that is, of the form

$$M_{ij} = g_{ij}(u)\frac{d}{dx} + b_{ijk}(u)u_x^k, \qquad \det(g_{ij}(u)) \neq 0. \qquad (2.70)$$

Let us introduce coefficients $\Gamma^i_{jk}(u)$ by the formula

$$b_{ijk}(u) = g_{is}(u)\Gamma^s_{jk}(u). \qquad (2.71)$$

Under local changes of coordinates $u^i = u^i(\tilde{u})$ on a manifold M the coefficient $g_{ij}(u)$ behaves as a metric on M, and the coefficient $\Gamma^s_{jk}(u)$ is transformed like the Christoffel symbols of an affine connection on M (see formulae (2.10) and (2.11) from Section 2.1.2 for the transformation law of symplectic operators):

$$\tilde{g}_{ij}(\tilde{u}) = \frac{\partial u^k}{\partial \tilde{u}^i} g_{ks}(u(\tilde{u})) \frac{\partial u^s}{\partial \tilde{u}^j}, \qquad (2.72)$$

$$\tilde{\Gamma}^i_{jk}(\tilde{u}) = \Gamma^p_{sr}(u(\tilde{u}))\frac{\partial \tilde{u}^i}{\partial u^p}\frac{\partial u^s}{\partial \tilde{u}^j}\frac{\partial u^r}{\partial \tilde{u}^k} + \frac{\partial \tilde{u}^i}{\partial u^p}\frac{\partial^2 u^p}{\partial \tilde{u}^j \partial \tilde{u}^k}. \qquad (2.73)$$

Lemma 2.2 ([109]) *Symplectic structures on the loop space ΩM defined by non-degenerate homogeneous symplectic operators of the first order (2.70) have the form*

$$\omega(\xi, \eta) = \int_\gamma \langle \xi, \nabla_{\dot\gamma}\eta \rangle, \qquad (2.74)$$

where $\xi = \xi(u(x))$, $\eta = \eta(u(x))$ are arbitrary smooth vector fields defined along the loop γ, $\langle \xi, \eta \rangle = g_{ij}(u(x))\xi^i(u(x))\eta^j(u(x))$ is the scalar product on $T_{\gamma(x)}M$ generated by the pseudo-Riemannian metric $g_{ij}(u)$, and $\nabla_{\dot\gamma}$ is a covariant differentiation along the loop $\gamma = \{u^i(x), 1 \le i \le N, x \in S^1\}$ on M defined by the affine connection $\Gamma^i_{jk}(u)$ and satisfying some additional conditions. All the forms (2.74) are invariant with respect to the group $\mathrm{Diff}^+(S^1)$ of diffeomorphisms of the circle preserving the orientation, that is, they do not depend on the parametrization of the loop γ.

Remark 2.4 *Note that if we consider arbitrary vector fields of the form $\xi = \xi(x, u, u_x, \ldots)$, then it is necessary to change everywhere the operators $\nabla_{\dot\gamma} = \nabla_{u_x}$ to the operators $\nabla_{u_x} - L_{u_x}$, where L_{u_x} is the Lie derivative along u_x, that is, $L_{u_x}\xi = [u_x, \xi]$.*

Theorem 2.1 ([109]) *Let (M, g_{ij}) be an arbitrary Riemannian or pseudo-Riemannian manifold. An affine connection $\Gamma^i_{jk}(u)$ on M defines on ΩM a symplectic structure (2.74) if and only if*

1. *the connection $\Gamma^i_{jk}(u)$ is compatible with the metric $g_{ij}(u)$, that is,*

$$\nabla_k g_{ij} \equiv \frac{\partial g_{ij}}{\partial u^k} - g_{is}\Gamma^s_{jk} - g_{js}\Gamma^s_{ik} = 0; \qquad (2.75)$$

2. *the torsion of the connection*

$$T_{ijk} = g_{is}T^s_{jk}, \qquad T^i_{jk} = \Gamma^i_{jk} - \Gamma^i_{kj}, \qquad (2.76)$$

is a closed 3-form on the manifold M.

Corollary 2.5 ([109]) *For an arbitrary pseudo-Riemannian manifold (M, g_{ij}), homogeneous closed 2-forms of the first order (2.74) on the loop space ΩM are in one-to-one correspondence with closed 3-forms on the manifold M.*

On any pseudo-Riemannian manifold M an arbitrary 3-form $T_{ijk}\, du^i \wedge du^j \wedge du^k$ always generates the unique connection $\Gamma^i_{jk}(u)$, which is compatible with the pseudo-Riemannian metric $g_{ij}(u)$ and has the torsion tensor $T_{ijk}(u)$ which coincides with this 3-form:

$$b_{sjk}(u) = g_{si}(u)\Gamma^i_{jk}(u) =$$

$$= \frac{1}{2}\left(\frac{\partial g_{sk}}{\partial u^j} + \frac{\partial g_{js}}{\partial u^k} - \frac{\partial g_{jk}}{\partial u^s} + T_{sjk}(u)\right). \qquad (2.77)$$

Thus, on the loop space of any Riemannian or pseudo-Riemannian manifold (M, g_{ij}) there are as many different homogeneous symplectic structures of the first order (2.74) as there are different closed 3-forms on the manifold M. This fact enabled the present author to assume that in this way it is possible to find the cohomology groups of the complexes of homogeneous forms on loop spaces of smooth manifolds via the de Rham cohomology groups of the manifolds. This idea was realized by the present author in the papers [114], [116] (see also [115]).

Corollary 2.6 ([109]) *For an arbitrary pseudo-Riemannian manifold (M, g_{ij}) a symplectic structure of the form (2.74) on ΩM is always defined. It is generated by the Levi-Civita connection (which is symmetric and compatible with the metric g_{ij}).*

A trivial consequence of Theorem 2.1 is the fact that a unique homogeneous symplectic structure of the first order is associated with an arbitrary two-dimensional pseudo-Riemannian metric, and therefore on any two-dimensional pseudo-Riemannian manifold (M, g_{ij}) there exists a unique symplectic structure of the form (2.74) on ΩM (it is generated by the Levi-Civita connection).

It is also obvious that symplectic (presymplectic) forms (2.74) are, generally speaking, degenerate on ΩM: the null space of the 2-forms (2.74) consists of parallel vector fields along the loop γ. And for any *point loop*, that is, a loop of the form $\gamma\colon S^1 \to P_0, \gamma(x) = \{u_0^i\}, x \in S^1$, where P_0 is an arbitrary point of the manifold M, $P_0 \in M$, any constant vector field along the point loop $\gamma\colon \xi(x) = \xi_0 = \text{const}, \xi_0 \in T_{P_0}M$, always belongs to the null space of the operator (2.70). The velocity vector field $\nu = \{u_x^i\}$ of the loop $\gamma = \{u^i(x)\}$ belongs to the null space of the 2-form (2.74) if and only if γ is a closed geodesic on the manifold M.

Theorem 2.1 gives a complete differential-geometric description of local non-degenerate homogeneous symplectic operators of the first order (2.70).

Locally, using the Poincaré lemma on exactness of closed forms in any coordinate neighbourhood, we deduce that all symplectic operators (2.70) for an arbitrary symmetric matrix function $g_{ij}(u)$ are defined, according to formulae (2.70) and (2.77), by an arbitrary skew-symmetric matrix function $f_{ij}(u)$:

$$f_{ij}(u) = -f_{ji}(u), \tag{2.78}$$

$$T_{ijk}(u) = \frac{\partial f_{ij}}{\partial u^k} + \frac{\partial f_{jk}}{\partial u^i} + \frac{\partial f_{ki}}{\partial u^j}. \tag{2.79}$$

Remark 2.5 Note that a complete explicit description or a classification of all local symplectic and Poisson structures of the first order has not yet been found. Only the case of scalar (one-component) operators, that is, the case $N = 1$, has been studied in detail. All local scalar symplectic operators of the first order have the form [22]:

$$I = 2p\frac{d}{dx} + p_x, \tag{2.80}$$

$$p = \frac{d}{dx}\left(\frac{\partial G}{\partial u_{xx}}\right) - \frac{\partial G}{\partial u_x}, \tag{2.81}$$

where $G = G(x, u, u_x, u_{xx})$ is an arbitrary function.

Local one-component Hamiltonian operators have been completely described up to the fifth order inclusive. In particular, all local one-component Hamiltonian operators of the first order have the form ([186], see also [103]):

$$L = \pm\frac{1}{\frac{\delta A}{\delta u(x)}}\frac{d}{dx} \circ \frac{1}{\frac{\delta A}{\delta u(x)}}, \tag{2.82}$$

where $A = \int a(x, u, u_x)\, dx$, and $a = a(x, u, u_x)$ is an arbitrary function. By means of contact transformations all Hamiltonian operators of the first order (2.82) can be reduced to the canonical form $\pm\frac{d}{dx}$ ([5], see also [6]).

Note that the one-component case $N = 1$ is essentially special, namely, in this case contact geometry effectively plays the key role. Problems connected with the contact theory of one-component Poisson structures and Hamiltonian equations are outside the framework of the present paper and they will be considered in a separate survey (see [5], [6], [15], [16], [101]–[103], [106], [107], [111], [120], [122]).

2.2.2   Finite-dimensional reductions of homogeneous symplectic structures of the first order

Let us consider the question of the connection of our symplectic structures   (2.74) on ΩM with finite-dimensional symplectic structures, that is, the question of reductions of the form (2.74) to finite-dimensional manifolds (this question was considered and studied for the first time in [109]).

Example 2.1 ([109]) For N-dimensional Riemannian manifolds (M, g_{ij}) whose geodesics are periodic and are of equal length, the geodesic flow generates a free action of the group $S^1 \cong \mathbf{R}/\mathbf{Z}$ on the bundle UM of unit tangent vectors, and therefore the factor space UM/S^1 is a $(2N - 2)$-dimensional manifold, namely, the manifold of geodesics CM (see, for example, [9]).

The symplectic structure on CM (the Reeb symplectic form [174], [175]) is given by the curvature form of the S^1-connection in the principal bundle UM over CM of unit tangent vectors of the manifold (M, g_{ij}) (see [174], [175], [187], [9]). The tangent space $T_\gamma CM$ of the manifold of geodesics CM at the point γ is isomorphic to the space of normal Jacobian fields along the geodesic γ on the manifold M.

In the present author's paper [109] the symplectic structures (2.74) on the loop spaces ΩM of such special Riemannian manifolds M were considered, and the reductions of these symplectic structures to the finite-dimensional manifolds CM of closed geodesics on M were studied. In particular, in [109] it is proved that the reduction of the symplectic structure (2.74) defined on the loop space ΩM of the manifold M by the Levi-Civita connection to the finite-dimensional subspace of normal Jacobian fields along the geodesic

γ coincides with the Reeb form, which is a closed non-degenerate 2-form defining a finite-dimensional symplectic structure on CM. The well-known examples of such special Riemannian manifolds, whose geodesics are periodic and are of equal length, are the following: Euclidean spheres, projective spaces $\mathbf{KP}^n$ ($\mathbf{K}$ is the field of real numbers $\mathbf{R}$, complex numbers $\mathbf{C}$, or the division algebra of quaternions $\mathbf{H}$), the Cayley plane $\mathbf{CaP}^2$ (all these manifolds are considered with the canonical Riemannian metrics and all of them are compact symmetric spaces of rank one), and also Zoll manifolds (see [9]).

2.2.3 Symplectic representation for an arbitrary two-dimensional non-linear sigma-model

Let us consider now the motion equations of the general two-dimensional non-linear sigma-models with torsion defined by actions of the form

$$S = \int \left(\frac{1}{2}(g_{ij}(u) + f_{ij}(u))u_x^i u_t^j + U(u) \right) dx \, dt, \quad (2.83)$$

$$g_{ij} = g_{ji}, \quad f_{ij} = -f_{ji},$$

where $g_{ij}(u)$ is a pseudo-Riemannian metric on a manifold M and $U(u)$ is an arbitrary function on M.

Proposition 2.4 ([109]) *For any action S of the form (2.83) the corresponding Lagrangian system $\delta S/\delta u^i(x) = 0$ always has a symplectic representation*

$$M_{ij}[u(x)]u_t^j = \frac{\delta U}{\delta u^i(x)}, \quad (2.84)$$

where

$$M_{ij}[u(x)] = g_{ij}(u(x))\frac{d}{dx} + \frac{1}{2}\left[\left(\frac{\partial g_{ki}}{\partial u^j} + \frac{\partial g_{ij}}{\partial u^k} - \frac{\partial g_{kj}}{\partial u^i} \right) + \right.$$

$$\left. + \left(\frac{\partial f_{ki}}{\partial u^j} + \frac{\partial f_{ij}}{\partial u^k} - \frac{\partial f_{kj}}{\partial u^i} \right) \right]u_x^k \quad (2.85)$$

is a homogeneous symplectic first-order operator defined by the formulae (2.70), (2.77)–(2.79).

Thus, locally, all homogeneous symplectic first-order operators (2.70) exactly define the symplectic structures of two-dimensional non-linear sigma-models with torsion (2.83).

2.2.4 Examples of bi-Hamiltonian representations for non-linear sigma-models

The natural question concerning the description and the study of Poisson and symplectic structures compatible with symplectic operators (2.85), which gives the possibility of constructing integrable bi-Hamiltonian two-dimensional non-linear sigma-models, was initiated as a special research programme by the present author (see [113], [123]). For certain known integrable two-dimensional non-linear sigma-models such Poisson structures compatible with (2.85) were found subsequently by Meshkov in [99]. Let us consider some simple examples of such bi-Hamiltonian integrable non-linear sigma-models.

Example 2.2 ([99]) Consider the action

$$S = \int \left(\frac{1}{2} \frac{u_x^1 u_t^2 + u_x^2 u_t^1}{u^1 u^2 + c} + k u^1 u^2 \right) dx\, dt, \qquad (2.86)$$

where c and k are arbitrary constants. In this case

$$(g_{ij}) = \frac{1}{u^1 u^2 + c} \begin{pmatrix} 0 & 1 \\ 1 & 0 \end{pmatrix}, \qquad U(u) = k u^1 u^2. \qquad (2.87)$$

Correspondingly, the first symplectic operator (2.85) for the action (2.86) has the form

$$A = (A_{ij}) = \frac{1}{u^1 u^2 + c} \begin{pmatrix} 0 & 1 \\ 1 & 0 \end{pmatrix} \frac{d}{dx} -$$
$$- \frac{1}{(u^1 u^2 + c)^2} \begin{pmatrix} 0 & u^1 u_x^2 \\ u^2 u_x^1 & 0 \end{pmatrix}. \qquad (2.88)$$

For the action (2.86) the second Hamiltonian operator B, which is compatible with (2.88) (with A^{-1}), is non-local and has the form:

$$B = (B^{ij}) = \begin{pmatrix} 0 & -(u^1 u^2 + c) \\ (u^1 u^2 + c) & 0 \end{pmatrix} +$$

$$+ \begin{pmatrix} u_x^1 \left(\frac{d}{dx} \right)^{-1} \circ u^1 & -u_x^1 \left(\frac{d}{dx} \right)^{-1} \circ u^2 \\ u_x^2 \left(\frac{d}{dx} \right)^{-1} \circ u^1 & -u_x^2 \left(\frac{d}{dx} \right)^{-1} \circ u^2 \end{pmatrix} + \quad (2.89)$$

$$+ \begin{pmatrix} u^1 \left(\frac{d}{dx} \right)^{-1} \circ u_x^1 & u^1 \left(\frac{d}{dx} \right)^{-1} \circ u_x^2 \\ -u^2 \left(\frac{d}{dx} \right)^{-1} \circ u_x^1 & -u^2 \left(\frac{d}{dx} \right)^{-1} \circ u_x^2 \end{pmatrix} .$$

The recursion operator $R_j^i = B^{is} A_{sj}$ corresponding to the constructed bi-Hamiltonian representation of the integrable non-linear sigma-model (2.86) has the form

$$R = (R_j^i) = \begin{pmatrix} -1 & 0 \\ 0 & 1 \end{pmatrix} \frac{d}{dx} + \quad (2.90)$$

$$+ \frac{1}{u^1 u^2 + c} \begin{pmatrix} u^1 u_x^2 & 2u^1 u_x^1 \\ -2u^2 u_x^2 & -u^2 u_x^1 \end{pmatrix} +$$

$$+ \begin{pmatrix} u_x^1 \left(\frac{d}{dx} \right)^{-1} \circ \frac{c u_x^2}{(u^1 u^2 + c)^2} & u_x^1 \left(\frac{d}{dx} \right)^{-1} \circ \frac{-c u_x^1}{(u^1 u^2 + c)^2} \\ u_x^2 \left(\frac{d}{dx} \right)^{-1} \circ \frac{c u_x^2}{(u^1 u^2 + c)^2} & u_x^2 \left(\frac{d}{dx} \right)^{-1} \circ \frac{-c u_x^1}{(u^1 u^2 + c)^2} \end{pmatrix} +$$

$$+ \begin{pmatrix} u^1 \left(\frac{d}{dx} \right)^{-1} \\ -u^2 \left(\frac{d}{dx} \right)^{-1} \end{pmatrix} \circ \begin{pmatrix} \frac{u^1 (u_x^2)^2}{(u^1 u^2 + c)^2} - \frac{u_{xx}^2}{u^1 u^2 + c} & \frac{u^2 (u_x^1)^2}{(u^1 u^2 + c)^2} - \frac{u_{xx}^1}{u^1 u^2 + c} \end{pmatrix} .$$

Example 2.3 ([99]) Consider the action

$$S = \int \left(\frac{1}{2} \frac{u_x^1 u_t^2 + u_x^2 u_t^1}{u^1 + u^2} + k(u^1 + u^2) \right) dx \, dt, \quad (2.91)$$

where k is an arbitrary constant. In this case

$$(g_{ij}) = \frac{1}{u^1 + u^2} \begin{pmatrix} 0 & 1 \\ 1 & 0 \end{pmatrix}, \qquad U(u) = k(u^1 + u^2). \quad (2.92)$$

The first Hamiltonian structure for the action (2.91) is defined by the symplectic operator (2.85) and has the form

$$A = (A_{ij}) = \frac{1}{u^1 + u^2} \begin{pmatrix} 0 & 1 \\ 1 & 0 \end{pmatrix} \frac{d}{dx} -$$

$$- \frac{1}{(u^1 + u^2)^2} \begin{pmatrix} 0 & u_x^2 \\ u_x^1 & 0 \end{pmatrix} . \quad (2.93)$$

The second Hamiltonian structure for the action (2.91) is defined by a non-local Hamiltonian operator B which is compatible with (2.93) (with A^{-1}):

$$B = (B^{ij}) = \begin{pmatrix} 0 & -(u^1 + u^2) \\ (u^1 + u^2) & 0 \end{pmatrix} +$$

$$+ \begin{pmatrix} u_x^1 & -u_x^1 \\ u_x^2 & -u_x^2 \end{pmatrix} \left(\frac{d}{dx} \right)^{-1} +$$

$$+ \left(\frac{d}{dx} \right)^{-1} \circ \begin{pmatrix} u_x^1 & u_x^2 \\ -u_x^1 & -u_x^2 \end{pmatrix}. \tag{2.94}$$

The recursion operator $R_j^i = B^{is} A_{sj}$ has the form

$$R = (R_j^i) = \begin{pmatrix} -1 & 0 \\ 0 & 1 \end{pmatrix} \frac{d}{dx} + \tag{2.95}$$

$$+ \frac{1}{u^1 + u^2} \begin{pmatrix} u_x^2 & 2u_x^1 \\ -2u_x^2 & -u_x^1 \end{pmatrix} +$$

$$+ \begin{pmatrix} -u_x^1 \left(\frac{d}{dx} \right)^{-1} \circ \frac{u_x^2}{(u^1+u^2)^2} & u_x^1 \left(\frac{d}{dx} \right)^{-1} \circ \frac{u_x^1}{(u^1+u^2)^2} \\ -u_x^2 \left(\frac{d}{dx} \right)^{-1} \circ \frac{u_x^2}{(u^1+u^2)^2} & u_x^2 \left(\frac{d}{dx} \right)^{-1} \circ \frac{u_x^1}{(u^1+u^2)^2} \end{pmatrix} +$$

$$+ \left(\frac{d}{dx} \right)^{-1} \circ \begin{pmatrix} \frac{(u_x^2)^2}{(u^1+u^2)^2} - \frac{u_{xx}^2}{u^1+u^2} & \frac{(u_x^1)^2}{(u^1+u^2)^2} - \frac{u_{xx}^1}{u^1+u^2} \\ -\frac{(u_x^2)^2}{(u^1+u^2)^2} + \frac{u_{xx}^2}{u^1+u^2} & -\frac{(u_x^1)^2}{(u^1+u^2)^2} + \frac{u_{xx}^1}{u^1+u^2} \end{pmatrix}.$$

In these examples the second Hamiltonian structures are defined by non-local Hamiltonian operators of the same differential-geometric type:

$$B = (B^{ij}) = g^{ij}(u) + a^i(u) \left(\frac{d}{dx} \right)^{-1} \circ b_k^j(u) u_x^k +$$

$$+ b_k^i(u) u_x^k \left(\frac{d}{dx} \right)^{-1} \circ a^j(u), \tag{2.96}$$

$$\det(g^{ij}(u)) \neq 0, \qquad g^{ij}(u) = -g^{ji}(u).$$

This class of non-local Hamiltonian structures of differential-geometric type is connected with the geometry of almost symplectic manifolds and deserves a particular study (it is especially important to find the structures that are compatible with symplectic operators of the form (2.85)).

In the one-component case (for $N = 1$) the local symplectic operator

$$M = \frac{d}{dx} \tag{2.97}$$

is compatible with non-local Hamiltonian operators of the form

$$K_{\pm} = \frac{d}{dx} \pm u_x \left(\frac{d}{dx} \right)^{-1} \circ u_x \tag{2.98}$$

and, according to the Lenard–Magri scheme, the compatible pair (M, K_+) generates the integrable hierarchy

$$u_t = (K_+ M)^n u_x,$$

where n is an arbitrary integer, of the sine-Gordon equation

$$u_{xt} = \sin u. \tag{2.99}$$

The potentialized modified Korteweg–de Vries equation

$$u_t = u_{xxx} + \frac{1}{2} u_x^3 \tag{2.100}$$

also belongs to this integrable hierarchy.

The compatible pair (M, K_-) generates the integrable hierarchy

$$u_t = (K_- M)^n u_x,$$

where n is an arbitrary integer, of the Liouville equation

$$u_{xt} = e^u. \tag{2.101}$$

The equation of Korteweg–de Vries type

$$u_t = u_{xxx} - \frac{1}{2} u_x^3 \tag{2.102}$$

also belongs to this integrable hierarchy.

In connection with these examples it is very interesting to describe and to study the differential geometry of compatible matrix analogues of differential-geometric type for the compatible pairs $(M, K_{\pm})$, that is, compatible pairs (M_{ij}, K^{ij}), where M_{ij} is a local homogeneous symplectic first-order operator (2.70) and K^{ij} is a non-local homogeneous Hamiltonian operator of the form

$$K^{ij} = g^{ij}(u)\frac{d}{dx} + b_k^{ij}(u)u_x^k + Cu_x^i \left(\frac{d}{dx}\right)^{-1} \circ u_x^j. \qquad (2.103)$$

This problem has not yet been solved, although the conditions that the operators (2.70) and (2.103) are, respectively, symplectic and Hamiltonian and very interesting differential geometry connected with these conditions have been completely studied. Therefore, in this interesting unsolved problem, it is necessary to study only the compatibility conditions, which must be also connected with a beautiful geometry. It turns out that Hamiltonian operators of the form (2.103) play a very important role in the theory of systems of hydrodynamic type, Whitham equations, generalized Heisenberg ferromagnets and others. The present author and Ferapontov (see [151]) proved that under the assumption that $\det(g^{ij}(u)) \neq 0$ the operator (2.103) is Hamiltonian if and only if it is defined by a pseudo-Riemannian metric $g^{ij}(u)$ of constant Riemannian curvature C (in this case, the coefficients $b_k^{ij}(u)$ are the coefficients of the Levi-Civita connection completely generated by this metric).

2.3   Symplectic and Poisson structures of degenerate Lagrangian systems

2.3.1   Symplectic representations for degenerate Lagrangian systems

Note that the Lagrangian of the action (2.83) is degenerate and in order to construct a Hamiltonian formalism it requires, generally speaking, the use of the Dirac procedure for systems with constraints ([18], see also [164], [165]). In fact, there exist explicit formulae for symplectic operators that correspond to degenerate Lagrangians (including those dependent on higher order derivatives of the fields $u^i(x)$ with respect to time t).

Actually, let us consider two-dimensional Lagrangian systems generated by actions of the form

$$S = \int \left(g_i(x, u, u_x, \ldots)u_t^i + h(x, u, u_x, \ldots)\right) dx\, dt, \qquad (2.104)$$

where g_i, h are arbitrary functions of the independent variable x, the fields $u^i(x,t)$, and finitely many their derivatives $u^i_{(n)} = \partial^n u^i / \partial x^n$ with respect to x.

The corresponding Lagrangian system $\delta S / \delta u^i(x) = 0$ has the form

$$M_{ij}[u(x)]u^j_t = \frac{\delta H}{\delta u^i(x)}, \qquad (2.105)$$

where

$$M_{ij}[u(x)] = \frac{\partial g_i}{\partial u^j_{(n)}} \frac{d^n}{dx^n} - (-1)^n \frac{d^n}{dx^n} \circ \frac{\partial g_j}{\partial u^i_{(n)}}, \qquad (2.106)$$

$$H = \int h \, dx.$$

Lemma 2.3 *The differential operator M_{ij} (2.106) is a symplectic operator and the bilinear form*

$$\omega(\xi, \eta) = \int_{S^1} \xi^i M_{ij}[u(x)]\eta^j \, dx, \qquad (2.107)$$

where $\xi, \eta \in T_\gamma \Omega M$, is a symplectic (presymplectic) form on the loop space ΩM.

Hence, we always have a symplectic representation (2.105) for Lagrangian systems generated by actions of the form (2.104).

Let us prove Lemma 2.3. Consider the 1-form $\alpha(\xi)$ on ΩM:

$$\alpha(\xi) = \int_{S^1} \xi^i(x) g_i(x, u, u_x, \ldots) \, dx.$$

Then for its differential $(d\alpha)(\xi, \eta)$ we obtain:

$$(d\alpha)(\xi, \eta) = \int_{S^1} \xi^i(x) \frac{\delta \alpha(\eta)}{\delta u^i(x)} \, dx - \int_{S^1} \eta^i(x) \frac{\delta \alpha(\xi)}{\delta u^i(x)} \, dx =$$

$$= \int_{S^1} \xi^i (-1)^k \frac{d^k}{dx^k} \left(\frac{\partial g_j}{\partial u^i_{(k)}} \eta^j \right) dx - \int_{S^1} \xi^i \frac{\partial g_i}{\partial u^j_{(k)}} \frac{d^k}{dx^k} (\eta^j) \, dx =$$

$$= - \int_{S^1} \xi^i M_{ij} \eta^j \, dx.$$

The exact 2-form $(d\alpha)(\xi, \eta)$ on ΩM is closed and generates a fortiori the symplectic operator M_{ij}.

Note that the symplectic operator (2.106) can also be obtained by applying the Dirac procedure to the degenerate Lagrangian of the action (2.104), but this way is considerably longer and more complicated technically.

Some special actions of the form (2.104) generate important non-linear equations of mathematical physics and field theory such as non-linear sigma-models, Monge–Ampère equations, Hamiltonian systems of hydrodynamic type, equations of associativity in two-dimensional topological field theory, the Korteweg–de Vries equation, the Hopf equation, the Krichever–Novikov equation, and many others. The corresponding symplectic representations are very useful and effective for the investigation of the integrability of non-linear systems and for the application of averaging procedures and perturbation theory.

2.3.2   Bi-Lagrangian systems

Some important integrable systems possess two (and even more) local and different but *compatible Lagrangian representations* (that is, the symplectic structures corresponding to them are compatible). We shall call systems of this type *bi-Lagrangian systems.* Thus, the class of bi-Lagrangian systems is a special case of the class of bi-Hamiltonian integrable systems. It would be very interesting to construct a general theory of systems possessing different Lagrangian representations and therefore obeying to different variational principles. It is also very interesting to study the conditions for Lagrangians possessing a "double" and to investigate nature of the compatibility of such Lagrangians.

Let us consider here some curious examples of bi-Lagrangian systems.

Example 2.4 *Bi-Lagrangian property of the Korteweg–de Vries equation (KdV).*

Let us consider the following action of the form (2.104):

$$S_1 = \int \left(\frac{1}{2} u_x u_t - \frac{1}{2} u_{xx}^2 - u_x^3 \right) dx\, dt. \qquad (2.108)$$

Symplectic representation for the Lagrangian equation

$$\frac{\delta S_1}{\delta u(x)} = 0,$$

that is,

$$u_t = 3u_x^2 - u_{xxx}, \tag{2.109}$$

has the form:

$$M_1(u_t) = -\frac{\delta H_1}{\delta u(x)},$$

$$M_1 = \frac{d}{dx}, \qquad H_1 = \int \left(u_x^3 + \frac{1}{2}u_{xx}^2 \right) dx. \tag{2.110}$$

The same equation (2.109) possesses a different Lagrangian representation with the action S_2 of the form (2.104):

$$S_2 = \int \left[\left(\frac{1}{2}u_{xxx} - u_x^2 \right) u_t + \frac{1}{2}u_{xxx}^2 + 5u_x u_{xx}^2 + \frac{5}{2}u_x^4 \right] dx\, dt. \tag{2.111}$$

The corresponding symplectic representation (2.105), (2.106) has the form:

$$M_2 = \frac{d^3}{dx^3} - 4u_x \frac{d}{dx} - 2u_{xx}, \tag{2.112}$$

$$H_2 = \int \left(\frac{1}{2}u_{xxx}^2 + \frac{5}{2}u_x^4 + 5u_x u_{xx}^2 \right) dx.$$

Moreover, the constructed local symplectic operators M_1 and M_2 are compatible. Two local symplectic structures for the potentialized KdV equation (2.109) were found in [22].

After the transformation $v(x) = u_x(x)$ we obtain the usual well-known Hamiltonian Gardner–Zakharov–Faddeev representation for the KdV equation:

$$v_t = \frac{d}{dx}\frac{\delta H_1}{\delta v(x)}, \qquad H_1 = \int \left(v^3 + \frac{1}{2}v_x^2 \right) dx, \tag{2.113}$$

$$v_t = 6vv_x - v_{xxx}. \tag{2.114}$$

Correspondingly, for the second symplectic structure we obtain the following formulae:

$$\left(\frac{d}{dx} - 2\left(\frac{d}{dx} \right)^{-1} \circ v - 2v\left(\frac{d}{dx} \right)^{-1} \right) v_t = -\frac{\delta H_2}{\delta v(x)}, \tag{2.115}$$

$$H_2 = \int \left(\frac{1}{2} v_{xx}^2 + 5 v v_x^2 + \frac{5}{2} v^4 \right) dx.$$

It is easy to describe all non-local matrix symplectic operators of the type of the symplectic structure (2.115) for the KdV equation that are given in the differential-geometric form invariant with respect to local changes of coordinates on a manifold, that is, the operators of the form (see [115], [117])

$$\begin{aligned}
M_{ij} &= g_{ij}(u) \frac{d}{dx} + b_{ijk}(u) u_x^k + \\
&+ a_i(u) \left(\frac{d}{dx} \right)^{-1} \circ b_j(u) + b_i(u) \left(\frac{d}{dx} \right)^{-1} \circ a_j(u).
\end{aligned}$$

The local part

$$g_{ij}(u) \frac{d}{dx} + b_{ijk}(u) u_x^k$$

is generated by an arbitrary homogeneous symplectic operator of the first order, which were completely described above, and the non-local part is defined by two arbitrary closed 1-forms $a_i(u) du^i$ and $b_i(u) du^i$ on the manifold M, that is,

$$\frac{\partial a_i}{\partial u^j} - \frac{\partial a_j}{\partial u^i} = 0, \qquad \frac{\partial b_i}{\partial u^j} - \frac{\partial b_j}{\partial u^i} = 0,$$

and, locally, there exist functions $f(u)$ and $h(u)$ such that

$$a_i(u) = \frac{\partial f}{\partial u^i}, \qquad b_i(u) = \frac{\partial h}{\partial u^i}.$$

The operators

$$K_{n-1} = R^{n-1} \frac{d}{dx}, \qquad (2.116)$$

where

$$R = \frac{d^2}{dx^2} - 4v - 2v_x \left(\frac{d}{dx} \right)^{-1} \qquad (2.117)$$

is the recursion operator for the KdV equation, for any integer n define compatible Hamiltonian structures for the KdV equation. Local Lagrangian representations (2.108) and (2.111) generate the well-known first and "zero" Hamiltonian structures (2.113) and (2.115) of the KdV equation corresponding to $n = 1, 0$. Magri's Hamiltonian operator (the second Hamiltonian structure of the KdV equation)

$$\pm \left[\frac{d^3}{dx^3} - 4\lambda v \frac{d}{dx} - 2\lambda v_x \right], \qquad \lambda = \text{const}, \qquad (2.118)$$

corresponds to $n = 2$. For $n = 3$ we obtain the third Hamiltonian structure of the KdV equation, namely, the non-local Hamiltonian fifth-order operator

$$M_5 = \frac{d^5}{dx^5} - 8v \frac{d^3}{dx^3} - 12v_x \frac{d^2}{dx^2} - 8v_{xx} \frac{d}{dx} + 16v^2 \frac{d}{dx} -$$

$$-2v_{xxx} + 16vv_x - 4v_x \left(\frac{d}{dx} \right)^{-1} \circ v_x. \qquad (2.119)$$

In the present author's paper [111] (see also [117]) a transformation connecting the Hamiltonian operator (2.119) with the canonical constant Gardner–Zakharov–Faddeev Hamiltonian operator $\frac{d}{dx}$ was constructed. Note that the transformation presented later in the paper [74] for the same purpose coincides precisely with the transformation found by the present author in [111] (see also [117]). This transformation is a higher analogue of the Miura transformation [98] taking, as is well known (see [60], [84]), Magri's Hamiltonian operator (2.118) to the canonical Gardner–Zakharov–Faddeev Hamiltonian operator $\frac{d}{dx}$.

Example 2.5 *Bi-Lagrangian property of the Hopf equation.*
    Let us consider the simplest example, namely, the Hopf equation

$$u_t = u_x^3, \qquad (2.120)$$

which possesses the following compatible Lagrangian representations with the actions S_1 and S_2 of the form (2.104):

$$S_1 = \int \left(\frac{1}{2} u_x u_t - \frac{1}{4} u_x^4 \right) dx\, dt, \qquad (2.121)$$

$$S_2 = \int \left(-\frac{u_t}{2u_x} - u_x^2 \right) dx\, dt. \qquad (2.122)$$

    Bisymplectic representation of the Hopf equation was demonstrated by Dorfman and the present author in [24].

54   O.I. MOKHOV

Example 2.6 *Lagrangian representation of the Krichever–Novikov equation (KN).*
Let us consider the following action of the form (2.104):

$$S_1 = \int \left(\frac{1}{2} \frac{u_t}{u_x} - \frac{1}{2} \frac{u_{xx}^2}{u_x^2} - \frac{1}{3} \frac{R(u)}{u_x^2} \right) dx\, dt, \qquad (2.123)$$

where $R(u) = a_3 u^3 + a_2 u^2 + a_1 u + a_0$ is an arbitrary third-order polynomial, $a_i = $ const, $i = 0, 1, 2, 3$. The corresponding Lagrangian equation $\delta S_1 / \delta u(x) = 0$ possesses the symplectic representation (2.105), (2.106) with the local symplectic operator

$$M_1 = -\frac{1}{u_x} \frac{d}{dx} \circ \frac{1}{u_x}, \qquad (2.124)$$

$$\left(\frac{1}{u_x} \frac{d}{dx} \circ \frac{1}{u_x} \right) u_t = \frac{\delta H}{\delta u(x)}, \qquad (2.125)$$

$$H = \int \left(\frac{1}{2} \frac{u_{xx}^2}{u_x^2} + \frac{1}{3} \frac{R(u)}{u_x^2} \right) dx.$$

Relations (2.125) define the well-known (Sokolov, [177]) non-local Hamiltonian and local symplectic representation for the Krichever–Novikov equation [81]:

$$u_t = u_{xxx} - \frac{3}{2} \frac{u_{xx}^2}{u_x} + \frac{R(u)}{u_x}. \qquad (2.126)$$

In the present author's paper [111] the canonical Hamiltonian representation for the Krichever–Novikov equation is constructed.

Example 2.7 *Bi-Lagrangian property of the Schwarzian–KdV equation (the degenerate Krichever–Novikov equation).*
Let us consider the case of the degenerate KN equation (2.126) with $R(u) \equiv 0$ (this equation is also known as the Schwarzian–KdV equation):

$$u_t = u_x S[u], \qquad (2.127)$$

where

$$S[u] = \frac{u_{xxx}}{u_x} - \frac{3}{2} \frac{u_{xx}^2}{u_x^2} \qquad (2.128)$$

is Schwarzian.

In this case, along with the Lagrangian representation generated by the action (2.123), where $R(u) \equiv 0$, equation (2.127) possesses one more local Lagrangian representation, which is also generated by an action of the form (2.104):

$$S_2 = \int \left[-\left(\frac{1}{2u_x} \right)_{xx} u_t - \frac{1}{2} u_x^{-2} u_{xxx}^2 + \frac{3}{8} u_x^{-4} u_{xx}^4 \right] dx \, dt. \quad (2.129)$$

The local symplectic representation corresponding to the action S_2 (2.129) is compatible with the local symplectic representation (2.125) for $R(u) \equiv 0$ and has the form:

$$M_2[u(x)]u_t = \frac{\delta H_2}{\delta u(x)},$$

$$M_2[u(x)] = \frac{1}{u_x^2} \frac{d^3}{dx^3} - 3 \frac{u_{xx}}{u_x^2} \frac{d^2}{dx^2} - \left(\frac{u_{xxx}}{u_x^3} - 3 \frac{u_{xx}^2}{u_x^4} \right) \frac{d}{dx}, \quad (2.130)$$

$$H_2 = \int \left(-\frac{1}{2} u_x^{-2} u_{xxx}^2 + \frac{3}{8} u_x^{-4} u_{xx}^4 \right) dx. \quad (2.131)$$

The second local symplectic structure for the equation (2.127) was found by Dorfman in [22].

2.3.3 Symplectic representation of the Monge–Ampère equation

Example 2.8 *The Monge–Ampère equation.*

Let us consider the following action of the form (2.104) [166], [167]:

$$S = \int \left(\frac{1}{2} u_x^2 q_t - u_x q_x u_t - \frac{1}{2} q^2 u_{xx} + \Phi(x, t, u, u_x) \right) dx \, dt, \quad (2.132)$$

where $\Phi(x, t, u, u_x)$ is an arbitrary function.

The local symplectic representation (2.105), (2.106) of the corresponding Lagrangian system $\delta S / \delta u^i(x) = 0$ has the form

$$M \begin{pmatrix} u \\ q \end{pmatrix}_t = \begin{pmatrix} \delta F / \delta u(x) \\ \delta F / \delta q(x) \end{pmatrix}, \quad (2.133)$$

where

$$M = \begin{pmatrix} q_x \frac{d}{dx} + \frac{d}{dx} \circ q_x & -u_{xx} \\ u_{xx} & 0 \end{pmatrix} \qquad (2.134)$$

is a symplectic operator, $u^1 = u$, $u^2 = q$, and

$$F = \int \left[\frac{1}{2} q^2 u_{xx} - \Phi(x, t, u, u_x) \right] dx. \qquad (2.135)$$

The symplectic system (2.133)–(2.135) is equivalent to the Monge–Ampère equation (see [167])

$$u_{xx} u_{tt} - (u_{xt})^2 = \frac{\delta \Phi}{\delta u(x)}. \qquad (2.136)$$

It is curious that in this case the corresponding Poisson structure is also local. It is defined by the Hamiltonian operator $K^{ij} = (M^{-1})^{ij}$:

$$K = \begin{pmatrix} 0 & \frac{1}{u_{xx}} \\ -\frac{1}{u_{xx}} & \frac{q_x}{u_{xx}^2} \frac{d}{dx} + \frac{d}{dx} \circ \frac{q_x}{u_{xx}^2} \end{pmatrix}. \qquad (2.137)$$

Simultaneous localization of symplectic and Hamiltonian operators inverse to each other remains valid for more general local Hamiltonian operators, namely, for arbitrary Hamiltonian operators of the form

$$K = \begin{pmatrix} 0 & B \\ -B & A\frac{d}{dx} + \frac{d}{dx} \circ A \end{pmatrix}, \qquad (2.138)$$

where B and A are arbitrary functions of the variables

$$x, u, u_x, u_{xx}, \ldots, q, q_x, q_{xx}, \ldots$$

($B \not\equiv 0$). In this case the corresponding symplectic operator has the form

$$M = \begin{pmatrix} \frac{A}{B^2} \frac{d}{dx} + \frac{d}{dx} \circ \frac{A}{B^2} & -\frac{1}{B} \\ \frac{1}{B} & 0 \end{pmatrix}. \qquad (2.139)$$

In particular, the operator

$$M = \begin{pmatrix} \left[\frac{\partial g}{\partial u_x} - \left(\frac{\partial g}{\partial u_{xx}} \right)_x \right] \frac{d}{dx} + \frac{d}{dx} \circ \left[\frac{\partial g}{\partial u_x} - \left(\frac{\partial g}{\partial u_{xx}} \right)_x \right] & \frac{\partial g}{\partial q} \\ -\frac{\partial g}{\partial q} & 0 \end{pmatrix}$$
$$(2.140)$$

is a symplectic operator of the form (2.139) for an arbitrary function $g(x, u, u_x, u_{xx}, q)$. The corresponding local Hamiltonian operator K has the form (2.138), where

$$B = -\frac{1}{(\partial g/\partial q)}, \qquad A = \frac{\frac{\partial g}{\partial u_x} - \left(\frac{\partial g}{\partial u_{xx}}\right)_x}{(\partial g/\partial q)^2}. \qquad (2.141)$$

The symplectic operator (2.140) can be obtained from formula (2.106), deduced in Lemma 2.3, for $g_1 = g(x, u, u_x, u_{xx}, q)$, $g_2 = 0$ ($u^1 = u$, $u^2 = q$).

To conclude this section we consider a more general case of degenerate Lagrangian systems defined by actions of the form:

$$S = \int \left(g_i(x, u, u_x, u_t, \ldots, u_{x^s t^r}, \ldots, u_{x^n t^m}) u^i_{t^{m+1}} + \right.$$

$$\left. + h(x, u, u_x, u_t, \ldots, u_{x^s t^r}, \ldots, u_{x^k t^m}) \right) dx \, dt, \qquad (2.142)$$

where $u_{x^s t^r} = \partial^{s+r} u/\partial x^s \partial t^r$, $r \leq m$. In other words, the Lagrangian is linear in regard to the higher derivatives of the fields $u^i(x, t)$ with respect to t.

Introducing the new variables $q^i_s(x, t) = u^i_{t^s}$, $0 \leq s \leq m$, we have the following relations for them: $(q^i_s)_t = q^i_{s+1}$, $0 \leq s \leq m-1$.

It is obvious that the Lagrangian system $\delta S/\delta u^i(x) = 0$ is equivalent to the Lagrangian system $\delta \widehat{S}/\delta q^i_s(x) = 0$, $\delta \widehat{S}/\delta \lambda^i_k(x) = 0$, $0 \leq k \leq m-1$, where

$$\widehat{S} = \int \left(\widehat{g}_i(x, \ldots, (q^r_s)_{x^k}, \ldots)(q^i_m)_t + q^i_s(\lambda^i_s)_t + \right.$$

$$\left. + \lambda^i_s q^i_{s+1} + \widehat{h}(x, \ldots, (q^r_s)_{x^k}, \ldots) \right) dx \, dt, \qquad (2.143)$$

and $\lambda^i_k(x, t)$ are Lagrange multipliers.

Thus, we obtain an action $\widehat{S}$ of the form (2.104) and have the corresponding symplectic representation (2.105), (2.106) for the Lagrangian system $\delta S/\delta u^i(x) = 0$.

2.4   Homogeneous symplectic structures of the second order on loop spaces of almost symplectic and symplectic manifolds and symplectic connections

2.4.1   General homogeneous symplectic forms of arbitrary orders

Let us consider homogeneous matrix differential symplectic operators $A_{ij}^{[m]}$ of the form

$$A_{ij}^{[m]} = a_{ij}^{[m]}(u)\frac{d^m}{dx^m} + b_{ijk}^{[m]}(u)u_x^k\frac{d^{m-1}}{dx^{m-1}} + \qquad (2.144)$$

$$+\left(c_{ijk}^{[m]}(u)u_{xx}^k + c_{ijkl}^{[m]}(u)u_x^k u_x^l\right)\frac{d^{m-2}}{dx^{m-2}} + \cdots + d_{ijk}^{[m]}(u)u_{(m)}^k + \cdots,$$

where each term has a homogeneity degree m with respect to the natural grading:

$$\deg(fg) = \deg f + \deg g, \qquad \deg f(u(x)) = \deg u(x) = 0,$$

$$\deg u_{(k)} = \deg \frac{d^k u}{dx^k} = \deg \frac{d^k}{dx^k} = k.$$

Definition 2.8 Symplectic operators of the form (2.144) are called *homogeneous symplectic operators of differential-geometric type.*

Homogeneous Poisson differential-geometric type structures defined by homogeneous differential operators of the form (2.144) (but with upper indices as is necessary for Hamiltonian operators) were considered for the first time by Dubrovin and Novikov in [35] (see also [33], [36]) where the problem of their classification was posed. It is obvious that the descriptions of homogeneous Hamiltonian and symplectic operators of the differential-geometric form (2.144) essentially differ and are connected with quite different differential-geometric structures on manifolds. In Section 4 we shall concentrate on the Dubrovin–Novikov homogeneous Poisson structures, their generalizations and applications.

Symplectic operators of the form (2.144) and the corresponding symplectic (presymplectic) forms on the loop spaces ΩM

$$\omega(\xi, \eta) = \int_{S^1} \xi^i (A_{ij}^{[m]} \eta^j) \, dx \qquad (2.145)$$

were introduced and completely studied for small m ($m = 1, 2$) by the present author in [109], [110], [124].

It is easy to show that any local changes of coordinates $u^i = u^i(\tilde{u})$ on the manifold M preserve the homogeneity and the form of the operator (2.144), and its coefficients are transformed as differential-geometric objects on the manifold (see the general formulae (2.10) and (2.11) for transformation of any symplectic operator under a local change of coordinates on a manifold in Section 2.1.2).

It follows from the skew-symmetry of the operator (2.144) that the higher coefficient $a_{ij}^{[m]}(u)$, which is always transformed as a metric on the manifold under local changes of coordinates, must be a symmetric matrix whenever m is odd, $m = 2k + 1$, and a skew-symmetric matrix whenever m is even, $m = 2k$.

If $\det(a_{ij}^{[m]}(u)) \neq 0$, then the corresponding symplectic forms can be expressed in an invariant form via the metric $a_{ij}^{[m]}(u)$ and the curvature and torsion tensors of the affine connections $a^{[m]is} b_{sjk}^{[m]}$, $a^{[m]is} c_{sjk}^{[m]}, \ldots, a^{[m]is} d_{sjk}^{[m]}$:

$$\omega(\xi, \eta) = \int_{S^1} \left(\langle \xi, \nabla_{\tilde{\gamma}}^m \eta \rangle + \cdots \right) dx,$$

where $\langle \xi, \eta \rangle = a_{ij}^{[m]}(u) \xi^i \eta^j$, $a^{[m]is} a_{sj}^{[m]} = \delta_j^i$, and here dots mean terms with lower degrees of the covariant derivative.

For $m = 0$ the symplectic conditions for the operator (2.144) have the form

$$a_{ij}^{[0]}(u) = -a_{ji}^{[0]}(u), \qquad \sum_{(i,j,k)} \frac{\partial a_{ij}^{[0]}}{\partial u^k} = 0,$$

where the sum is taken over all cyclic permutations of i, j, k.

This means that

$$\Omega_0 = a_{ij}^{[0]}(u) \, du^i \wedge du^j$$

is a closed 2-form on the manifold M and

$$\omega(\xi, \eta) = \int_{S^1} \Omega_0(\xi, \eta)\, dx$$

is an arbitrary ultralocal symplectic structure on the loop space ΩM.

Thus, the case $m = 0$ corresponds exactly to classic finite-dimensional symplectic geometry.

The case $m = 1$ corresponds to the homogeneous symplectic operators of the first order (2.70) and symplectic forms (2.74) from Section 2.2.

2.4.2 Homogeneous symplectic forms of the second order

Consider the case $m = 2$, that is, symplectic operators of the form

$$A_{ij}^{[2]} = a_{ij}^{[2]}(u)\frac{d^2}{dx^2} + b_{ijk}^{[2]}(u)u_x^k\frac{d}{dx} +$$
$$+c_{ijk}^{[2]}(u)u_{xx}^k + c_{ijkl}^{[2]}(u)u_x^k u_x^l,\quad \det(a_{ij}^{[2]}(u)) \neq 0. \quad (2.146)$$

In this case, $g_{ij}(u) = a_{ij}^{[2]}(u)$ is a non-degenerate skew-symmetric tensor field of type $(0,2)$ on M, that is, the manifold M is *almost symplectic*, and $g_{ij}(u)$ is the *almost symplectic structure* on M.

Let us introduce the coefficients $\Gamma_{jk}^i(u)$ by the formula

$$b_{ijk}^{[2]}(u) = 2g_{is}(u)\Gamma_{jk}^s(u). \quad (2.147)$$

Later in this section, we omit the symbol [2] in the coefficients of the homogeneous operator (2.146) of order 2.

Lemma 2.4 ([110]) *The coefficients $\Gamma_{jk}^i(u)$ define a symplectic connection on (M, g_{ij}), that is, an affine connection which is compatible with the almost symplectic structure $g_{ij}(u)$ on the manifold:*

$$\nabla_k g_{ij} \equiv \frac{\partial g_{ij}}{\partial u^k} - g_{is}\Gamma_{jk}^s - g_{sj}\Gamma_{ik}^s = 0. \quad (2.148)$$

Lemma 2.4 is a direct consequence of the conditions of skew-symmetry for the homogeneous operator (2.146). In fact, the operator (2.146) is skew-symmetric if and only if the following relations for the coefficients are fulfilled:

$$g_{ij} = -g_{ji}, \tag{2.149}$$

$$\frac{\partial g_{ij}}{\partial u^k} = \frac{1}{2}(b_{ijk} - b_{jik}), \tag{2.150}$$

$$\sum_{(k,l)}\sum_{(i,j)} \left(2c_{ijkl} - \frac{\partial b_{ijk}}{\partial u^l}\right) = 0, \tag{2.151}$$

$$\sum_{(i,j)} \left(c_{ijk} - \frac{1}{2}b_{ijk}\right) = 0, \tag{2.152}$$

where $\sum_{(i,j)}$ means summation over all permutations of the corresponding elements (i,j) in parentheses under the summation sign.

Relation (2.150) corresponds exactly to the condition that the coefficients $\Gamma^i_{jk}(u)$ define a symplectic connection on (M, g_{ij}).

It turns out that any symplectic connection $\Gamma^i_{jk}(u)$ on the given almost symplectic manifold (M, g_{ij}) generates a symplectic operator of the form (2.146), and, moreover, for any fixed almost symplectic structure $g_{ij}(u)$ any symplectic connection uniquely defines the corresponding symplectic operator (2.146), that is, all other coefficients of the symplectic operator (2.146) are uniquely defined by the metric $g_{ij}(u)$ and the connection $\Gamma^i_{jk}(u)$.

Theorem 2.2 ([110]) *If (M, g_{ij}) is an almost symplectic manifold, then there is a one-to-one correspondence between symplectic connections $\Gamma^i_{jk}(u)$ on (M, g_{ij}) and closed 2-forms (2.145), (2.146), $m = 2$, on the loop space ΩM. This correspondence can be expressed by the following formula for arbitrary homogeneous closed 2-forms (2.145), (2.146), $m = 2$, of the second order:*

$$\omega(\xi, \eta) = \int_{S^1} \left\{ \langle \xi, \nabla^2_\nu \eta \rangle + \langle T(\xi, \eta), \nabla_\nu \nu \rangle + \frac{1}{2} \langle \nu, R(\xi, \eta)\nu \rangle + \right.$$

$$\left. + \left\langle \nu, \sum_{(\eta, \xi, \nu)} R(\eta, \xi)\nu \right\rangle + \langle T(\nu, \xi), T(\nu, \eta) \rangle \right\} dx, \tag{2.153}$$

where $\sum_{(\eta,\xi,\nu)}$ means summation over all cyclic permutations of
the elements (η,ξ,ν); $\xi(u(x))$ and $\eta(u(x))$ are arbitrary smooth
vector fields along the loop $\gamma = \{u^i(x), 1 \leq i \leq N, x \in S^1\}$, $\langle \xi, \eta \rangle =
g_{ij}(u(x))\xi^i(u(x))\eta^j(u(x))$ is the scalar product on the tangent space
of the almost symplectic manifold (M, g_{ij}), and the vector field ν is
the velocity vector field of the loop $\gamma(x)$, that is, $\nu^i = u_x^i$; ∇_ν is the
covariant differentiation along the loop γ defined by an arbitrary
symplectic connection $\Gamma_{jk}^i(u)$ on (M, g_{ij});

$$[T(\xi, \eta)]^i = T_{kl}^i \xi^k \eta^l, \qquad T_{jk}^i \equiv \Gamma_{jk}^i - \Gamma_{kj}^i,$$

is the torsion tensor of the symplectic connection;

$$[R(\xi, \eta)\zeta]^i = R_{jkl}^i \xi^k \eta^l \zeta^j,$$

$$R_{jkl}^i = -\frac{\partial \Gamma_{jl}^i}{\partial u^k} + \frac{\partial \Gamma_{jk}^i}{\partial u^l} - \Gamma_{pk}^i \Gamma_{jl}^p + \Gamma_{pl}^i \Gamma_{jk}^p,$$

is the curvature tensor of the symplectic connection.

Formula (2.153) gives the general form of homogeneous symplectic structures of the second order (2.145), (2.146), $m = 2$, on
the loop space ΩM and also describes all symplectic operators of
the form (2.146).

Proof. If the relations of skew-symmetry (2.149)–(2.152) are
fulfilled, then the closedness of the corresponding 2-form is equivalent to the following relations:

$$c_{ijk} = \frac{1}{2}(b_{ijk} + b_{kji} - b_{kij}), \qquad (2.154)$$

$$c_{ijkl} + c_{ijlk} = \frac{\partial c_{ijk}}{\partial u^l} + \frac{\partial c_{ijl}}{\partial u^k} + \frac{\partial c_{lik}}{\partial u^j} - \frac{\partial c_{ljk}}{\partial u^i}. \qquad (2.155)$$

It is easy to verify that the relation (2.152) is a direct corollary
of the relation (2.154), and the relation (2.151) is a corollary of the
relations (2.154) and (2.155).

Thus, the operator (2.146) is symplectic if and only if the relations (2.149), (2.150), (2.154), and (2.155) are fulfilled, whence
it follows that for an arbitrary almost symplectic structure $g_{ij}(u)$
all symplectic operators of the form (2.146), $a_{ij}^{[2]}(u) = g_{ij}(u)$, are

uniquely defined by arbitrary symplectic connections $\Gamma^i_{jk}(u)$ on the almost symplectic manifold (M, g_{ij}).

It is important to note that the class of symplectic connections on almost symplectic manifolds is very rich. It is easy to show that on each almost symplectic manifold there exist infinitely many different symplectic connections and, correspondingly, infinitely many different symplectic forms (2.153) on its loop space.

Let us find now explicit expressions for the coefficients of the symplectic operator (2.146) via invariant differential-geometric objects on the manifold M.

For the corresponding covariant derivative ∇_ν of a vector field $\xi(u(x))$ along a loop $(\nu^k = u^k_x)$, we have:

$$(\nabla_\nu \xi)^i = \nu^k \frac{\partial \xi^i}{\partial u^k} + \nu^k \Gamma^i_{jk}(u)\xi^j(u) = \frac{d\xi^i}{dx} + u^k_x \Gamma^i_{jk}(u)\xi^j(u),$$

$$(\nabla^2_\nu \xi)^i = \frac{d^2\xi^i}{dx^2} + \nu^k_x \Gamma^i_{jk}\xi^j + \nu^k \nu^l \frac{\partial \Gamma^i_{jl}}{\partial u^k}\xi^j + 2\nu^k \Gamma^i_{jk}\frac{d\xi^j}{dx} + \nu^l \nu^k \Gamma^i_{pl}\Gamma^p_{jk}\xi^j,$$

$$
\begin{aligned}
g_{ij}(\nabla^2_\nu \xi)^j &= g_{ij}\frac{d^2\xi^j}{dx^2} + 2g_{is}u^k_x \Gamma^s_{jk}\frac{d\xi^j}{dx} + \\
&+ g_{is}u^k_x u^l_x \left(\frac{\partial \Gamma^s_{jl}}{\partial u^k} + \Gamma^s_{pk}\Gamma^p_{jl}\right)\xi^j + g_{is}u^k_{xx}\Gamma^s_{jk}\xi^j,
\end{aligned}
$$

$$c_{ijk}u^k_{xx} = g_{is}\Gamma^s_{jk}u^k_{xx} + T_{kji}u^k_{xx},$$

$$(\nabla_\nu \nu)^k = u^k_{xx} + \Gamma^k_{il}u^i_x u^l_x,$$

$$c_{ijk}u^k_{xx} = T_{kji}(\nabla_\nu \nu)^k + g_{is}\Gamma^s_{jk}u^k_{xx} - \frac{1}{2}(T_{sji}\Gamma^s_{kl} + T_{sji}\Gamma^s_{lk})u^k_x u^l_x,$$

$$
\begin{aligned}
\frac{\partial c_{ijk}}{\partial u^l} &= \frac{\partial}{\partial u^l}(g_{is}\Gamma^s_{jk} + T_{kji}) = \\
&= g_{ir}\Gamma^r_{sl}\Gamma^s_{jk} + g_{rs}\Gamma^r_{il}\Gamma^s_{jk} + g_{is}\frac{\partial \Gamma^s_{jk}}{\partial u^l} + \frac{\partial T_{kji}}{\partial u^l},
\end{aligned}
$$

where we use the relation (2.148), namely, the compatibility of the symplectic connection with the almost symplectic structure.

Further we have:

$$c_{ijkl} + c_{ijlk} = g_{ir}\Gamma^r_{sl}\Gamma^s_{jk} + g_{rs}\Gamma^r_{il}\Gamma^s_{jk} + g_{is}\frac{\partial \Gamma^s_{jk}}{\partial u^l} + \frac{\partial T_{kji}}{\partial u^l} +$$

$$+ g_{ir}\Gamma^r_{sk}\Gamma^s_{jl} + g_{rs}\Gamma^r_{ik}\Gamma^s_{jl} + g_{is}\frac{\partial \Gamma^s_{jl}}{\partial u^k} + \frac{\partial T_{lji}}{\partial u^k} +$$

$$+ g_{lr}\Gamma^r_{sj}\Gamma^s_{ik} + g_{rs}\Gamma^r_{lj}\Gamma^s_{ik} + g_{ls}\frac{\partial \Gamma^s_{ik}}{\partial u^j} + \frac{\partial T_{kil}}{\partial u^j} -$$

$$- g_{lr}\Gamma^r_{si}\Gamma^s_{jk} - g_{rs}\Gamma^r_{li}\Gamma^s_{jk} - g_{ls}\frac{\partial \Gamma^s_{jk}}{\partial u^i} - \frac{\partial T_{kjl}}{\partial u^i}.$$

Thus, for the symplectic operator (2.146), we obtain:

$$A^{[2]}_{ij}\xi^j = g_{ij}(\nabla^2_\nu \xi)^j + T_{kji}(\nabla_\nu \nu)^k \xi^j + \frac{1}{2}F_{ijkl}(u)u^k_x u^l_x \xi^j,$$

where

$$F_{ijkl}(u) = g_{rs}\Gamma^r_{il}\Gamma^s_{jk} + \frac{\partial T_{kji}}{\partial u^l} + g_{rs}\Gamma^r_{ik}\Gamma^s_{jl} + \frac{\partial T_{lji}}{\partial u^k} +$$

$$+ g_{lr}\Gamma^r_{sj}\Gamma^s_{ik} + g_{rs}\Gamma^r_{lj}\Gamma^s_{ik} + g_{ls}\frac{\partial \Gamma^s_{ik}}{\partial u^j} + \frac{\partial T_{kil}}{\partial u^j} - g_{lr}\Gamma^r_{si}\Gamma^s_{jk} -$$

$$- g_{rs}\Gamma^r_{li}\Gamma^s_{jk} - g_{ls}\frac{\partial \Gamma^s_{jk}}{\partial u^i} - \frac{\partial T_{kjl}}{\partial u^i} - T_{sji}\Gamma^s_{kl} - T_{sji}\Gamma^s_{lk}.$$

We transform $F_{ijkl}(u)$ using necessary identities of almost symplectic geometry:

$$F_{ijkl}(u) = g_{ls}\frac{\partial T^s_{ik}}{\partial u^j} - g_{ls}\frac{\partial T^s_{jk}}{\partial u^i} - g_{ls}R^s_{kji} + g_{ls}\Gamma^s_{pj}T^p_{ik} -$$

$$- g_{ls}\Gamma^s_{pi}T^p_{jk} + g_{rs}\Gamma^s_{jk}T^r_{il} + g_{rs}\Gamma^s_{ik}T^r_{lj} + \frac{\partial T_{kji}}{\partial u^l} +$$

$$+ \frac{\partial T_{lji}}{\partial u^k} + \frac{\partial T_{kil}}{\partial u^j} + \frac{\partial T_{klj}}{\partial u^i} - T_{sji}\Gamma^s_{kl} - T_{sji}\Gamma^s_{lk} =$$

$$= \frac{\partial T_{lik}}{\partial u^j} - g_{ps}\Gamma^p_{lj}T^s_{ik} - \frac{\partial T_{ljk}}{\partial u^i} + g_{ps}\Gamma^p_{li}T^s_{jk} - R_{lkji} +$$

$$+ g_{rs}\Gamma^s_{jk}T^r_{il} + g_{rs}\Gamma^s_{ik}T^r_{lj} + \frac{\partial T_{kji}}{\partial u^l} + \frac{\partial T_{lji}}{\partial u^k} +$$

$$+\frac{\partial T_{kil}}{\partial u^j} + \frac{\partial T_{klj}}{\partial u^i} - T_{sji}\Gamma^s_{kl} - T_{sji}\Gamma^s_{lk} =$$

$$= -R_{lkji} + \sum_{(i,k,j)} \frac{\partial T_{lik}}{\partial u^j} + \sum_{(i,l,j)} \frac{\partial T_{kil}}{\partial u^j} - g_{ps}\Gamma^p_{lj}T^s_{ik} +$$

$$+ g_{ps}\Gamma^p_{li}T^s_{jk} + g_{rs}\Gamma^s_{jk}T^r_{il} + g_{rs}\Gamma^s_{ik}T^r_{lj} - T_{sji}\Gamma^s_{kl} - T_{sji}\Gamma^s_{lk} =$$

$$= R_{lkij} + \sum_{(i,k,j)} (\nabla_j T_{lik} + T_{lpj}T^p_{ki}) + \sum_{(i,l,j)} (\nabla_j T_{kil} + T_{kpj}T^p_{li}) +$$

$$+ T^p_{kj}T_{pil} + T^p_{ki}T_{plj}.$$

The identity

$$\sum_{(j,k,l)} (R^i_{jkl} - \nabla_j T^i_{kl} + T^i_{pl}T^p_{jk}) = 0$$

is always fulfilled for any affine connection. Therefore, for any affine connection compatible with a metric, we always have the following identity (the first Bianchi identity):

$$\sum_{(j,k,l)} (R_{ijkl} - \nabla_j T_{ikl} + T_{ipl}T^p_{jk}) = 0.$$

Using the last udentity, we obtain:

$$F_{ijkl}(u) = R_{lkij} + \sum_{(i,k,j)} R_{likj} + \sum_{(i,l,j)} R_{kilj} + T^p_{kj}T_{pil} + T^p_{ki}T_{plj}.$$

Correspondingly, for the symplectic operator, we obtain the following explicit expression invariant with respect to local changes of coordinates on the manifold M:

$$A^{[2]}_{ij}\xi^j = g_{ij}(\nabla^2_\nu\xi)^j + T_{kji}(\nabla_\nu\nu)^k\xi^j +$$

$$+ \left[\frac{1}{2}R_{lkij} + \sum_{(i,k,j)} R_{likj} + T^p_{kj}T_{pil}\right]u^k_x u^l_x \xi^j,$$

and we obtain the invariant formula (2.153) for the corresponding symplectic structure $\omega(\xi, \eta)$.

We note that for any symplectic connection the corresponding curvature tensor is symmetric with respect to the pair of the first indices:

$$R_{ijkl} = R_{jikl}$$

(we recall that curvature tensors corresponding to Riemannian connections compatible with Riemannian or pseudo-Riemannian metrics are skew-symmetric with respect to the pair of the first indices: $R_{ijkl} = -R_{jikl}$).

On symplectic manifolds there exists a special class of symmetric symplectic connections satisfying the condition $T^i_{jk} = 0$. It is easy to show that symmetric symplectic connections exist on (M, g_{ij}) if and only if (M, g_{ij}) is a symplectic manifold, that is, $(dg)_{ijk} = 0$.

Actually, let $\Gamma^i_{jk}(u)$ be a symplectic connection, that is,

$$\frac{\partial g_{ij}}{\partial u^k} = g_{is}\Gamma^s_{jk} + g_{sj}\Gamma^s_{ik},$$

where $g_{ij}(u)$ is a skew-symmetric non-degenerate matrix function (a metric) on M.

By permuting the indices (ijk) in this relation cyclically and defining $\Gamma_{ijk} = g_{is}\Gamma^s_{jk}$, $T_{ijk} = g_{is}T^s_{jk}$, $T^i_{jk} = \Gamma^i_{jk} - \Gamma^i_{kj}$, we obtain

$$\frac{\partial g_{ij}}{\partial u^k} = \Gamma_{ijk} - \Gamma_{jik},$$

$$\frac{\partial g_{jk}}{\partial u^i} = \Gamma_{jki} - \Gamma_{kji},$$

$$\frac{\partial g_{ki}}{\partial u^j} = \Gamma_{kij} - \Gamma_{ikj}.$$

Summing all these relations, we obtain that all torsion tensors of symplectic connections always satisfy the identity

$$\sum_{(i,j,k)} T_{ijk} = \sum_{(i,j,k)} \frac{\partial g_{ij}}{\partial u^k}.$$

Therefore, if a symplectic connection is symmetric ($T^i_{jk} = 0$), then the almost symplectic structure g_{ij} gives a non-degenerate closed 2-form on the manifold M, that is, the manifold (M, g_{ij}) is symplectic.

We note that for any symmetric symplectic connection the following identities are always fulfilled:

$$R_{ijkl} + R_{iklj} + R_{iljk} = 0,$$

$$R_{ijkl} + R_{iklj} = R_{klij} - R_{ljik}.$$

Corollary 2.7 ([**110**]) *If the symplectic connection $\Gamma^i_{jk}(u)$ is symmetric, that is, $T^i_{jk}(u) = 0$, then the corresponding homogeneous symplectic form of the second order (2.153) on ΩM has the form*

$$\omega(\xi, \eta) = \int_{S^1} \left\{ \langle \xi, \nabla^2_\nu \eta \rangle + \frac{1}{2} \langle \nu, R(\xi, \eta)\nu \rangle \right\} dx. \qquad (2.156)$$

Formula (2.156) describes a natural class of homogeneous symplectic forms of the second order on loop spaces of symplectic manifolds.

On each symplectic manifold there exist infinitely many different symmetric symplectic connections, and consequently infinitely many corresponding symplectic forms (2.156) on ΩM.

3 Complexes of homogeneous forms on loop spaces of smooth manifolds and their cohomology groups

3.1 Homogeneous forms on loop spaces of smooth manifolds

The complexes $(\Omega_{[m]}, d)$ of homogeneous forms of an arbitrary fixed order m on loop spaces of smooth manifolds were introduced and studied in the present author's papers [114], [116].

Definition 3.1 ([114], [116]) By *homogeneous k-forms of order* m on the loop space ΩM of a smooth manifold M we mean skew-symmetric k-forms on ΩM of the following form at any point $\gamma \in \Omega M$:

$$\omega(\xi_1, \ldots, \xi_k) = \qquad\qquad\qquad\qquad\qquad (3.1)$$

$$= \int_{S^1} \left(\sum_1 M^{[n_1 \cdots n_k]}_{i_1 \cdots i_k}(u)(\xi_1^{i_1})_{(n_1)} \cdots (\xi_k^{i_k})_{(n_k)} + \right.$$

$$\left. + \sum_2 M^{[n_1 \cdots n_k p_1 \cdots p_s]}_{i_1 \cdots i_k j_1 \cdots j_s}(u) u^{j_1}_{(p_1)} \cdots u^{j_s}_{(p_s)} (\xi_1^{i_1})_{(n_1)} \cdots (\xi_k^{i_k})_{(n_k)} \right) dx,$$

where the first summation is taken over all $n_1, \ldots, n_k$ such that $n_1 + \cdots + n_k = m$, $n_r \geq 0$, and the second summation is taken over all $s > 0$ and all $n_1, \ldots, n_k, p_1, \ldots, p_s$, such that $n_1 + \cdots + n_k + p_1 + \cdots + p_s = m$, $n_r \geq 0$, $p_l > 0$, and, besides, the summations are taken over all $1 \leq i_r \leq N$, $1 \leq r \leq k$, $1 \leq j_l \leq N$, $1 \leq l \leq s$; $M^{[n_1 \cdots n_k p_1 \cdots p_s]}_{i_1 \cdots i_k j_1 \cdots j_s}(u)$ and $M^{[n_1 \cdots n_k]}_{i_1 \cdots i_k}(u)$ are arbitrary smooth functions of local coordinates on the manifold M, and $\xi_r(x) \in T_\gamma \Omega M$ are arbitrary smooth vector fields along the loop $\gamma(x) = \{u^i(x)\}$, $x \in S^1$.

We use here the standard notation for the total nth derivatives with respect to the independent variable x: $f_{(n)} = d^n f / dx^n$.

We can always assume without loss of generality that $0 < p_1 \leq \cdots \leq p_s$, and, moreover, that $n_1 = 0$ (provided that $k > 0$), since in our case $\int_{S^1}(dG/dx)dx = 0$ for all the relevant smooth functions $G(x, u(x), u_x(x), \ldots)$ defined on the loops γ.

It is obvious that the *set* $\Omega_{[m]}^k$ *of all homogeneous k-forms of order m on the loop space* ΩM is a linear space over **R**. The definition of the linear space $\Omega_{[m]}^k$ does not depend on local coordinates $\{u^i\}$, since local changes of coordinates $u^i = u^i(v^1, ..., v^N)$ on the manifold M preserve the homogeneity of k-forms and leave homogeneous k-forms of order m in the same class of homogeneity. The number m is an invariant of the homogeneous k-forms of order m with respect to local changes of coordinates on the manifold.

We shall also use the following notation: $\Omega_{[m]}$ *is the linear space of all homogeneous skew-symmetric forms of order m on the loop space* ΩM *of the manifold* M, Ω^k *is the linear space of all homogeneous skew-symmetric k-forms on* ΩM, *and* Ω *is the linear space of all homogeneous skew-symmetric forms on* ΩM:

$$\Omega = \sum_{m \geq 0} \Omega_{[m]} = \sum_{k \geq 0} \Omega^k = \sum_{m \geq 0} \sum_{k \geq 0} \Omega_{[m]}^k.$$

3.2 Complexes of homogeneous forms on loop spaces of smooth manifolds

We define the *differential d* on homogeneous k-forms by the following formula:

$$(d\omega)(\xi_1, ..., \xi_{k+1}) = \tag{3.2}$$

$$= \sum_{q=1}^{k+1} (-1)^{q+1} \int_{S^1} (\xi_q^i)_{(r)} (\xi_1^{i_1})_{(n_1)} \cdots$$

$$\cdots (\xi_{q-1}^{i_{q-1}})_{(n_{q-1})} (\xi_{q+1}^{i_q})_{(n_q)} \cdots (\xi_{k+1}^{i_k})_{(n_k)} \times$$

$$\times \frac{\partial \left(M_{i_1 \cdots i_k}^{[n_1 \cdots n_k]}(u) + M_{i_1 \cdots i_k j_1 \cdots j_s}^{[n_1 \cdots n_k p_1 \cdots p_s]}(u) u_{(p_1)}^{j_1} \cdots u_{(p_s)}^{j_s} \right)}{\partial u_{(r)}^i} dx.$$

Proposition 3.1 ([114], [116]) *The differential d is a linear map on the linear space* Ω *of homogeneous forms and it takes a homogeneous k-form of order m to a homogeneous* $(k + 1)$-*form of the same order m:*

$$d: \Omega_{[m]}^k \longrightarrow \Omega_{[m]}^{k+1},$$

with $d^2 = 0$.

Proof.

$$(d^2\omega)(\xi_1, ..., \xi_{k+2}) =$$

$$= \sum_{a=1}^{k+2} (-1)^{a+1} \int_{S^1} (\xi_a^t)_{(c)} \left[\sum_{q=1}^{a-1} (-1)^{q+1} (\xi_q^i)_{(r)} (\xi_1^{i_1})_{(n_1)} \cdots \right.$$

$$\cdots (\xi_{q-1}^{i_{q-1}})_{(n_{q-1})} (\xi_{q+1}^{i_q})_{(n_q)} \cdots$$

$$\cdots (\xi_{a-1}^{i_{a-2}})_{(n_{a-2})} (\xi_{a+1}^{i_{a-1}})_{(n_{a-1})} \cdots (\xi_{k+2}^{i_k})_{(n_k)} \times$$

$$\times \frac{\partial^2 \left(M_{i_1 \cdots i_k j_1 \cdots j_s}^{[n_1 \cdots n_k p_1 \cdots p_s]}(u) u_{(p_1)}^{j_1} \cdots u_{(p_s)}^{j_s} \right)}{\partial u_{(r)}^i \partial u_{(c)}^t} dx +$$

$$+ \sum_{q=a}^{k+1} (-1)^{q+1} (\xi_{q+1}^i)_{(r)} (\xi_1^{i_1})_{(n_1)} \cdots (\xi_{a-1}^{i_{a-1}})_{(n_{a-1})} (\xi_{a+1}^{i_a})_{(n_a)} \cdots$$

$$\cdots (\xi_q^{i_{q-1}})_{(n_{q-1})} (\xi_{q+2}^{i_q})_{(n_q)} \cdots (\xi_{k+2}^{i_k})_{(n_k)} \times$$

$$\times \frac{\partial^2 \left(M_{i_1 \cdots i_k j_1 \cdots j_s}^{[n_1 \cdots n_k p_1 \cdots p_s]}(u) u_{(p_1)}^{j_1} \cdots u_{(p_s)}^{j_s} \right)}{\partial u_{(r)}^i \partial u_{(c)}^t} dx \right] =$$

$$= \sum_{a=1}^{k+2} (-1)^{a+1} \int_{S^1} (\xi_a^t)_{(c)} \left[\sum_{q<a} (-1)^{q+1} (\xi_q^i)_{(r)} (\xi_1^{i_1})_{(n_1)} \cdots \right.$$

$$\cdots (\xi_{q-1}^{i_{q-1}})_{(n_{q-1})} (\xi_{q+1}^{i_q})_{(n_q)} \cdots$$

$$\cdots (\xi_{a-1}^{i_{a-2}})_{(n_{a-2})} (\xi_{a+1}^{i_{a-1}})_{(n_{a-1})} \cdots (\xi_{k+2}^{i_k})_{(n_k)} \times$$

$$\times \frac{\partial^2 \left(M_{i_1 \cdots i_k j_1 \cdots j_s}^{[n_1 \cdots n_k p_1 \cdots p_s]}(u) u_{(p_1)}^{j_1} \cdots u_{(p_s)}^{j_s} \right)}{\partial u_{(r)}^i \partial u_{(c)}^t} dx -$$

$$- \sum_{q=a+1}^{k+2} (-1)^{q+1} (\xi_q^i)_{(r)} (\xi_1^{i_1})_{(n_1)} \cdots (\xi_{a-1}^{i_{a-1}})_{(n_{a-1})} (\xi_{a+1}^{i_a})_{(n_a)} \cdots$$

$$\cdots (\xi_{q-1}^{i_{q-2}})_{(n_{q-2})} (\xi_{q+1}^{i_{q-1}})_{(n_{q-1})} \cdots (\xi_{k+2}^{i_k})_{(n_k)} \times$$

$$\times \frac{\partial^2 \left(M_{i_1 \cdots i_k j_1 \cdots j_s}^{[n_1 \cdots n_k p_1 \cdots p_s]}(u) u_{(p_1)}^{j_1} \cdots u_{(p_s)}^{j_s} \right)}{\partial u_{(r)}^i \partial u_{(c)}^t} dx \right] =$$

$$= \int_{S^1} \left[\sum_{a=1}^{k+2} \sum_{q<a} (-1)^{q+1} (-1)^{a+1} (\xi_a^t)_{(c)} (\xi_q^i)_{(r)} (\xi_1^{i_1})_{(n_1)} \cdots \right.$$

$$\cdots (\xi_{q-1}^{i_{q-1}})_{(n_{q-1})}(\xi_{q+1}^{i_q})_{(n_q)} \cdots$$

$$\cdots (\xi_{a-1}^{i_{a-2}})_{(n_{a-2})}(\xi_{a+1}^{i_{a-1}})_{(n_{a-1})} \cdots (\xi_{k+2}^{i_k})_{(n_k)} \times$$

$$\times \frac{\partial^2 \left(M_{i_1 \cdots i_k j_1 \cdots j_s}^{[n_1 \cdots n_k p_1 \cdots p_s]}(u) u_{(p_1)}^{j_1} \cdots u_{(p_s)}^{j_s} \right)}{\partial u_{(r)}^i \, \partial u_{(c)}^t} -$$

$$- \sum_{q=1}^{k+2} \sum_{q<a} (-1)^{q+1}(-1)^{a+1}(\xi_q^t)_{(c)}(\xi_a^i)_{(r)}(\xi_1^{i_1})_{(n_1)} \cdots$$

$$\cdots (\xi_{q-1}^{i_{q-1}})_{(n_{q-1})}(\xi_{q+1}^{i_q})_{(n_q)} \cdots$$

$$\cdots (\xi_{a-1}^{i_{a-2}})_{(n_{a-2})}(\xi_{a+1}^{i_{a-1}})_{(n_{a-1})} \cdots (\xi_{k+2}^{i_k})_{(n_k)} \times$$

$$\times \frac{\partial^2 \left(M_{i_1 \cdots i_k j_1 \cdots j_s}^{[n_1 \cdots n_k p_1 \cdots p_s]}(u) u_{(p_1)}^{j_1} \cdots u_{(p_s)}^{j_s} \right)}{\partial u_{(r)}^i \, \partial u_{(c)}^t} \Bigg] dx = 0.$$

Thus, we have constructed the *complex* (Ω, d) *of homogeneous skew-symmetric forms on the loop space* ΩM:

$$0 \xrightarrow{d} \Omega^0 \xrightarrow{d} \Omega^1 \xrightarrow{d} \Omega^2 \xrightarrow{d} \cdots. \qquad (3.3)$$

Moreover, for any m we have also constructed the *complex* $(\Omega_{[m]}, d)$ *of homogeneous skew-symmetric forms of order* m *on the loop space* ΩM:

$$0 \xrightarrow{d} \Omega_{[m]}^0 \xrightarrow{d} \Omega_{[m]}^1 \xrightarrow{d} \Omega_{[m]}^2 \xrightarrow{d} \cdots. \qquad (3.4)$$

Let $Z_{[m]}^k(\Omega M)$ be the linear space of all closed homogeneous k-forms ω of order m on ΩM ($d\omega = 0$), and let $B_{[m]}^k(\Omega M)$ be the linear subspace of all exact homogeneous k-forms ω of the order m on ΩM ($\omega = d\alpha$). We define the cohomology groups of our complex (3.4):

$$H_{[m]}^k(\Omega M, \mathbf{R}) = Z_{[m]}^k(\Omega M)/B_{[m]}^k(\Omega M). \qquad (3.5)$$

Definition 3.2 ([114], [116]) We shall call the groups

$$H_{[m]}^k(\Omega M, \mathbf{R})$$

the *homogeneous cohomology groups of order* m *of the loop space* ΩM *of the smooth manifold* M.

We also denote by $Z^k(\Omega M)$ the linear space of all closed homogeneous k-forms on ΩM and by $B^k(\Omega M)$ the linear space of all exact homogeneous k-forms on ΩM and define the *total homogeneous cohomology groups* $H^k(\Omega M, \mathbf{R})$ *of the loop space* ΩM *of the smooth manifold* M by

$$H^k(\Omega M, \mathbf{R}) = Z^k(\Omega M)/B^k(\Omega M). \qquad (3.6)$$

3.3 Cohomology groups of complexes of homogeneous forms on loop spaces of smooth manifolds

The zero-order homogeneous cohomology of the loop space ΩM coincides with the de Rham cohomology of the manifold M:

$$H^i_{[0]}(\Omega M, \mathbf{R}) = H^i(M, \mathbf{R}). \qquad (3.7)$$

Indeed, homogeneous skew-symmetric k-forms of order zero can be written as

$$\omega(\xi_1, ..., \xi_k) = \int_{S^1} M_{i_1 \cdots i_k}(u) \xi_1^{i_1} \cdots \xi_k^{i_k} dx, \qquad (3.8)$$

where $M_{i_1 \cdots i_k}(u)$ is an arbitrary smooth skew-symmetric covariant tensor field on the manifold M, and in this case the differential d (formula (3.2)) acts on these tensor fields as the classical de Rham differential:

$$(d\omega)(\xi_1, ..., \xi_{k+1}) =$$

$$= \sum_{q=1}^{k+1} (-1)^{q+1} \int_{S^1} \xi_q^i \xi_1^{i_1} \cdots \xi_{q-1}^{i_{q-1}} \xi_{q+1}^{i_q} \cdots \xi_{k+1}^{i_k} \frac{\partial M_{i_1 \cdots i_k}(u)}{\partial u^i} dx =$$

$$= \int_{S^1} (dM)_{i_1 \cdots i_{k+1}}(u) \xi_1^{i_1} \cdots \xi_{k+1}^{i_{k+1}} dx,$$

$$(dM)_{i_1 \cdots i_{k+1}}(u) = \sum_{q=1}^{k+1} (-1)^{q+1} \frac{\partial M_{i_1 \cdots \widehat{i_q} \cdots i_{k+1}}}{\partial u^{i_q}}.$$

Thus, our complex (3.3) is a very natural generalization of the classical de Rham complex on a smooth manifold to the case of loop spaces of smooth manifolds.

Apparently, all cohomology groups of the complex (3.3) of homogeneous forms on the loop space of a smooth manifold can be expressed explicitly via the de Rham cohomology groups of the manifold M.

Here we shall consider in detail the calculation of the first few homogeneous cohomology groups for the first non-trivial case $m = 1$. In particular, in this case the following theorem holds on the connection between the first-order homogeneous cohomology groups of the loop space of an arbitrary smooth manifold and the de Rham cohomology groups of the manifold.

Theorem 3.1 ([114], [116]) *For* $i = 0, 1, 2$

$$H^i_{[1]}(\Omega M, \mathbf{R}) = H^{i+1}(M, \mathbf{R}). \tag{3.9}$$

Proof. Let us consider arbitrary *homogeneous k-forms of the first order* on ΩM:

$$\omega(\xi_1, ..., \xi_k) = \int_{S^1} \left(\sum_{s=2}^k \xi_1^{i_1} M^{[s]}_{i_1 \cdots i_k}(u) \xi_2^{i_2} \cdots (\xi_s^{i_s})_x \cdots \xi_k^{i_k} + \right.$$
$$\left. + \xi_1^{i_1} M_{i_1 \cdots i_k j}(u) u_x^j \xi_2^{i_2} \cdots \xi_k^{i_k} \right) dx, \tag{3.10}$$

where $M^{[s]}_{i_1 \cdots i_k}(u)$ and $M_{i_1 \cdots i_k j}(u)$ are arbitrary smooth functions of local coordinates on the manifold M satisfying the conditions that ensure the skew-symmetry of the k-form (3.10).

Correspondingly, the first-order homogeneous 0-forms are the functionals F on ΩM that have the following form:

$$F = \int_{S^1} f_i(u) u_x^i dx, \tag{3.11}$$

where $f_i(u)$ is an arbitrary smooth covector field on the manifold M. Furthermore, note that in the case of loop space, the integral along any loop on the manifold of a total derivative with respect to the independent variable x is always equal to 0:

$$\int_{S^1} \left(\frac{d}{dx} f(u) \right) dx = \int_{S^1} \frac{\partial f}{\partial u^i} u_x^i dx = 0, \qquad (3.12)$$

where $f(u)$ is an arbitrary function on the manifold.

Let us find a criterion for a first-order homogeneous 0-form (3.11) to be closed. We have

$$(dF)(\xi) \equiv \int_{S^1} \xi^i \left(\frac{\partial f_k}{\partial u^i} - \frac{\partial f_i}{\partial u^k} \right) u_x^k dx, \qquad (3.13)$$

and, therefore, the 0-form (3.11) is closed, that is, $dF = 0$, if and only if

$$\frac{\partial f_k}{\partial u^i} - \frac{\partial f_i}{\partial u^k} = 0. \qquad (3.14)$$

The exact first-order homogeneous 0-forms are functionals of the form (3.11) that are equal to 0 identically, that is, by virtue of (3.12), 0-forms (3.11) such that

$$f_i = \frac{\partial f(u)}{\partial u^i}. \qquad (3.15)$$

Relation (3.14) means that the 1-form $f_i(u)du^i$ on the manifold M is closed, and relation (3.15) means that the 1-form $f_i(u)du^i$ on the manifold is exact.

Thus, we have proved that

$$H_{[1]}^0(\Omega M, \mathbf{R}) = H^1(M, \mathbf{R}).$$

Arbitrary first-order homogeneous 1-forms on the loop space have the following form:

$$\omega(\xi) = \int_{S^1} \xi^i f_{ik}(u) u_x^k dx, \qquad (3.16)$$

where $f_{ik}(u)$ is an arbitrary smooth covariant two-valent tensor field on the manifold M.

Let us find conditions for a first-order homogeneous 1-form (3.16) to be closed. We have

$$(d\omega)(\xi, \eta) = \int_{S^1} \xi^i \left(\eta^j \frac{\partial f_{jk}}{\partial u^i} u^k_x - (\eta^j f_{ji})_x \right) dx - \qquad (3.17)$$

$$- \int_{S^1} \eta^j \left(\xi^i \frac{\partial f_{ik}}{\partial u^j} u^k_x - (\xi^i f_{ij})_x \right) dx =$$

$$= \int_{S^1} \xi^i \left[\eta^j \left(\frac{\partial f_{jk}}{\partial u^i} - \frac{\partial f_{ji}}{\partial u^k} - \frac{\partial f_{ik}}{\partial u^j} \right) u^k_x - \eta^j_x (f_{ji} + f_{ij}) \right] dx.$$

Thus, a first-order homogeneous 1-form (3.16) is closed, that is, $d\omega = 0$, if and only if the following relations are valid:

$$f_{ji} + f_{ij} = 0, \qquad (3.18)$$

$$\frac{\partial f_{jk}}{\partial u^i} - \frac{\partial f_{ji}}{\partial u^k} - \frac{\partial f_{ik}}{\partial u^j} = 0. \qquad (3.19)$$

Relations (3.18) and (3.19) just mean that $f_{ij} du^i \wedge du^j$ is a closed 2-form on the manifold M. From formula (3.13) we find that the homogeneous 1-form (3.16) is exact if and only if

$$f_{ik} = \frac{\partial f_k}{\partial u^i} - \frac{\partial f_i}{\partial u^k}, \qquad (3.20)$$

that is, provided that $f_{ij} du^i \wedge du^j$ is an exact 2-form on the manifold M.

Thus, we have proved that

$$H^1_{[1]}(\Omega M, \mathbf{R}) = H^2(M, \mathbf{R}).$$

Arbitrary first-order homogeneous 2-forms have the expression

$$\omega(\xi, \eta) = \int_{S^1} \xi^i (g_{ij}(u)(\eta^j)_x + b_{ijk}(u) u^k_x \eta^j) dx, \qquad (3.21)$$

where $g_{ij}(u)$ is an arbitrary smooth covariant two-valent tensor field on the manifold M and $b_{ijk}(u)$ is an arbitrary smooth field on the manifold M whose components transform according to the following law under local changes of coordinates $u^i = u^i(v^1, ..., v^N)$ on the manifold:

$$\tilde{b}_{prs}(v) = b_{ijk}(u(v)) \frac{\partial u^i}{\partial v^p} \frac{\partial u^j}{\partial v^r} \frac{\partial u^k}{\partial v^s} + g_{ij}(u(v)) \frac{\partial u^i}{\partial v^p} \frac{\partial^2 u^j}{\partial v^r \partial v^s}. \qquad (3.22)$$

In particular, if $g_{ij}(u)$ is a non-degenerate symmetric metric on the manifold (in other words, if we deal with an arbitrary pseudo-Riemannian manifold (M, g_{ij})), then the coefficients $b_{ijk}(u)$ define an affine connection $\Gamma^i_{jk}(u)$ on the manifold M:

$$b_{ijk}(u) = g_{is}(u)\Gamma^s_{jk}(u), \qquad (3.23)$$

$$\tilde{\Gamma}^p_{rs}(v) = \Gamma^l_{jk}(u(v))\frac{\partial u^j}{\partial v^r}\frac{\partial u^k}{\partial v^s}\frac{\partial v^p}{\partial u^l} + \frac{\partial v^p}{\partial u^l}\frac{\partial^2 u^l}{\partial v^r \partial v^s}.$$

Moreover, the 2-form (3.21) must satisfy the skew-symmetry conditions

$$\omega(\xi, \eta) = -\omega(\eta, \xi),$$

that is,

$$\omega(\xi, \eta) + \omega(\eta, \xi) =$$
$$= \int_{S^1} [\xi^i g_{ij}(\eta^j)_x + \eta^j g_{ji}(\xi^i)_x + \xi^i(b_{ijk} + b_{jik})u^k_x \eta^j]dx =$$
$$= \int_{S^1} \xi^i \left[(g_{ij} - g_{ji})(\eta^j)_x + \left(b_{ijk} + b_{jik} - \frac{\partial g_{ji}}{\partial u^k}\right)u^k_x \eta^j\right]dx = 0,$$

whence

$$g_{ij} = g_{ji}, \qquad (3.24)$$

$$\frac{\partial g_{ij}}{\partial u^k} = b_{ijk} + b_{jik}. \qquad (3.25)$$

Note that for the most practically important case in which $g_{ij}(u)$ is a pseudo-Riemannian metric on the manifold M, relation (3.25) just means that the connection defined by (3.23) is compatible with the metric $g_{ij}(u)$:

$$\nabla_k g_{ij} \equiv \frac{\partial g_{ij}}{\partial u^k} - g_{is}\Gamma^s_{jk} - g_{sj}\Gamma^s_{ik} = 0.$$

Let us find the conditions for the 2-form (3.21) to be closed:

$$(d\omega)(\xi,\eta,\zeta) =$$

$$= \int_{S^1} \left[\xi^i \left(\frac{\partial g_{jk}}{\partial u^i} \eta^j (\zeta^k)_x + \frac{\partial b_{jks}}{\partial u^i} u_x^s \eta^j \zeta^k \right) + (\xi^i)_x b_{jki} \eta^j \zeta^k - \right.$$

$$-\eta^j \left(\frac{\partial g_{ik}}{\partial u^j} \xi^i (\zeta^k)_x + \frac{\partial b_{iks}}{\partial u^j} u_x^s \xi^i \zeta^k \right) - (\eta^j)_x b_{ikj} \xi^i \zeta^k +$$

$$\left. +\zeta^k \left(\frac{\partial g_{ij}}{\partial u^k} \xi^i (\eta^j)_x + \frac{\partial b_{ijs}}{\partial u^k} u_x^s \xi^i \eta^j \right) + (\zeta^k)_x b_{ijk} \xi^i \eta^j \right] dx =$$

$$= \int_{S^1} \xi^i \left[\left(\frac{\partial g_{ij}}{\partial u^k} - b_{jki} - b_{ikj} \right) (\eta^j)_x \zeta^k + \right.$$

$$+ \left(\frac{\partial g_{jk}}{\partial u^i} - b_{jki} - \frac{\partial g_{ik}}{\partial u^j} + b_{ijk} \right) \eta^j (\zeta^k)_x +$$

$$\left. + \left(\frac{\partial b_{jks}}{\partial u^i} - \frac{\partial b_{jki}}{\partial u^s} + \frac{\partial b_{ijs}}{\partial u^k} - \frac{\partial b_{iks}}{\partial u^j} \right) u_x^s \eta^j \zeta^k \right] dx \equiv 0.$$

Therefore, a first-order homogeneous skew-symmetric 2-form (3.21) is closed if and only if the following relations are valid (here we also assume that the skew-symmetry conditions (3.24) and (3.25) are satisfied):

$$\frac{\partial g_{ij}}{\partial u^k} = b_{ikj} + b_{jki}, \tag{3.26}$$

$$\frac{\partial b_{jks}}{\partial u^i} - \frac{\partial b_{jki}}{\partial u^s} + \frac{\partial b_{ijs}}{\partial u^k} - \frac{\partial b_{iks}}{\partial u^j} = 0. \tag{3.27}$$

The following theorem clarifies the differential-geometric meaning of relations (3.26) and (3.27) for skew-symmetric first-order homogeneous 2-forms (3.21) (that is, provided that relations (3.24) and (3.25) are satisfied).

Theorem 3.2 ([116]) *A first-order homogeneous 2-form (3.21) is closed if and only if $g_{ij} = g_{ji}$ (this condition is necessary for skew-symmetry) and*

$$b_{ijk} = \frac{1}{2} \left(\frac{\partial g_{ik}}{\partial u^j} + \frac{\partial g_{ji}}{\partial u^k} - \frac{\partial g_{jk}}{\partial u^i} + T_{ijk}(u) \right), \tag{3.28}$$

where $T_{ijk}(u)$ is the skew-symmetric tensor defined by an arbitrary closed 3-form $\alpha = T_{ijk}(u) du^i \wedge du^j \wedge du^k$ on the manifold M. Moreover, the 2-form (3.21) is exact if and only if α is an arbitrary exact

3-form on the manifold M, that is, $T_{ijk}(u)$ is the tensor field defined by an arbitrary skew-symmetric 2-form $\beta = \beta_{ij}(u)du^i \wedge du^j$ on the manifold M by the formula $T_{ijk} = (d\beta)_{ijk}$.

Proof. Let us introduce the functions $T_{ijk}(u)$ by the formula

$$T_{ijk} = b_{ijk} - b_{ikj}. \tag{3.29}$$

It follows from the transformation law (3.22) for the coefficients $b_{ijk}(u)$ that $T_{ijk}(u)$ is a tensor on M.

In particular, if $g_{ij}(u)$ is a non-degenerate pseudo-Riemannian metric, then $T_{ijk}(u)$ is the covariant torsion tensor with lower indices for the affine connection $\Gamma^s_{jk}(u)$ defined by the relation (3.23):

$$T_{ijk} = g_{is}T^s_{jk} = g_{is}(\Gamma^s_{jk} - \Gamma^s_{kj}).$$

By comparing (3.25) with (3.26), we obtain the relation

$$T_{ijk} = T_{jki}; \tag{3.30}$$

that is, since the tensor $T_{ijk}(u)$ is always skew-symmetric with respect to the last two indices by virtue of definition (3.29), we have proved that relation (3.26) for skew-symmetric 2-forms (3.21) is equivalent to the condition that the tensor $T_{ijk}(u)$ is completely skew-symmetric, that is, this tensor is skew-symmetric with respect to any pair of indices. Hence, the tensor $T_{ijk}(u)$ defines the skew-symmetric 3-form $\alpha = T_{ijk}(u)du^i \wedge du^j \wedge du^k$ on the manifold M.

Let us prove now that relation (3.27) is equivalent to the condition that the 3-form α defined by the skew-symmetric tensor $T_{ijk}(u)$ is closed on the manifold M.

We introduce the new functions

$$S_{ijsk}(u) = \frac{\partial b_{ijs}}{\partial u^k} - \frac{\partial b_{ijk}}{\partial u^s}. \tag{3.31}$$

By virtue of their definition, the functions $S_{ijsk}(u)$ are skew-symmetric with respect to the last two indices:

$$S_{ijsk} = -S_{ijks}. \tag{3.32}$$

Moreover, using formula (3.25), we find that the functions S_{ijsk} are also skew-symmetric with respect to the first two indices:

$$S_{ijsk} = \frac{\partial b_{ijs}}{\partial u^k} - \frac{\partial b_{ijk}}{\partial u^s} = -\frac{\partial b_{jis}}{\partial u^k} + \frac{\partial^2 g_{ij}}{\partial u^s \partial u^k} +$$

$$+ \frac{\partial b_{jik}}{\partial u^s} - \frac{\partial^2 g_{ij}}{\partial u^s \partial u^k} = -S_{jisk}. \tag{3.33}$$

Let us now derive a number of necessary auxiliary relations:

$$S_{ijkm} + S_{ikmj} + S_{imjk} = \frac{\partial b_{ijk}}{\partial u^m} - \frac{\partial b_{ijm}}{\partial u^k} + \frac{\partial b_{ikm}}{\partial u^j} -$$

$$- \frac{\partial b_{ikj}}{\partial u^m} + \frac{\partial b_{imj}}{\partial u^k} - \frac{\partial b_{imk}}{\partial u^j} =$$

$$= \frac{\partial T_{ijk}}{\partial u^m} - \frac{\partial T_{ijm}}{\partial u^k} + \frac{\partial T_{ikm}}{\partial u^j}. \tag{3.34}$$

Using the skew-symmetry properties (3.32) and (3.33) for $S_{ijkm}(u)$ and the skew-symmetry of the tensor $T_{ijk}(u)$, from (3.34) we obtain

$$2S_{ijkm} - 2S_{mkji} =$$

$$= S_{ijkm} + S_{ikmj} + S_{imjk} + S_{ijkm} + S_{kjmi} + S_{jmki} -$$

$$- S_{jmki} - S_{imjk} - S_{mkji} - S_{kjmi} - S_{ikmj} - S_{mkji} =$$

$$= S_{ijkm} + S_{ikmj} + S_{imjk} + S_{jimk} + S_{jkim} + S_{jmki} -$$

$$- S_{mjik} - S_{mikj} - S_{mkji} - S_{kjmi} - S_{kijm} - S_{kmij} =$$

$$= \frac{\partial T_{ijk}}{\partial u^m} - \frac{\partial T_{ijm}}{\partial u^k} + \frac{\partial T_{ikm}}{\partial u^j} + \frac{\partial T_{jim}}{\partial u^k} - \frac{\partial T_{jik}}{\partial u^m} + \frac{\partial T_{jmk}}{\partial u^i} -$$

$$- \frac{\partial T_{mji}}{\partial u^k} + \frac{\partial T_{mjk}}{\partial u^i} - \frac{\partial T_{mik}}{\partial u^j} - \frac{\partial T_{kjm}}{\partial u^i} + \frac{\partial T_{kji}}{\partial u^m} - \frac{\partial T_{kmi}}{\partial u^j} =$$

$$= \frac{\partial T_{jim}}{\partial u^k} + \frac{\partial T_{jki}}{\partial u^m} - \frac{\partial T_{mkj}}{\partial u^i} - \frac{\partial T_{kmi}}{\partial u^j}. \tag{3.35}$$

Relation (3.27) can be rewritten in the form

$$S_{jksi} - S_{iskj} + \frac{\partial T_{ijs}}{\partial u^k} - \frac{\partial T_{iks}}{\partial u^j} = 0. \tag{3.36}$$

Using formula (3.35), we further transform relation (3.27):

$$0 = S_{ijkm} - S_{mkji} + \frac{\partial T_{mik}}{\partial u^j} - \frac{\partial T_{mjk}}{\partial u^i} =$$

$$= \frac{1}{2} \left(\frac{\partial T_{jim}}{\partial u^k} + \frac{\partial T_{jki}}{\partial u^m} - \frac{\partial T_{mkj}}{\partial u^i} - \frac{\partial T_{kmi}}{\partial u^j} \right) + \frac{\partial T_{mik}}{\partial u^j} - \frac{\partial T_{mjk}}{\partial u^i} =$$

$$= \frac{1}{2} \left(\frac{\partial T_{jim}}{\partial u^k} + \frac{\partial T_{jki}}{\partial u^m} + \frac{\partial T_{mkj}}{\partial u^i} + \frac{\partial T_{kmi}}{\partial u^j} \right). \tag{3.37}$$

Thus, we have proved that relation (3.27) is equivalent to the condition that the 3-form α defined by the skew-symmetric tensor $T_{ijk}(u)$ is closed on the manifold M:

$$(dT)_{ijkm} \equiv \frac{\partial T_{jkm}}{\partial u^i} - \frac{\partial T_{ikm}}{\partial u^j} + \frac{\partial T_{ijm}}{\partial u^k} - \frac{\partial T_{ijk}}{\partial u^m} = 0, \qquad (3.38)$$

or $d\alpha = 0$.

By permuting indices in formula (3.25), we obtain

$$\frac{\partial g_{jk}}{\partial u^i} = b_{jki} + b_{kji}, \qquad (3.39)$$

$$\frac{\partial g_{ki}}{\partial u^j} = b_{kij} + b_{ikj}. \qquad (3.40)$$

Let us now consider the following linear combination of these formulae: (3.40) + (3.39) − (3.25). We obtain

$$2b_{kij} + T_{kji} + T_{jki} + T_{ikj} = -\frac{\partial g_{ij}}{\partial u^k} + \frac{\partial g_{jk}}{\partial u^i} + \frac{\partial g_{ki}}{\partial u^j}, \qquad (3.41)$$

and therefore,

$$b_{kij} = \frac{1}{2}\left(\frac{\partial g_{kj}}{\partial u^i} + \frac{\partial g_{ik}}{\partial u^j} - \frac{\partial g_{ij}}{\partial u^k} + T_{kij} \right). \qquad (3.42)$$

Conversely, let $g_{ij}(u)$ be an arbitrary symmetric tensor, and let $T_{ijk}(u)$ be the skew-symmetric tensor defined by an arbitrary closed 3-form $\alpha = T_{ijk}(u)du^i \wedge du^j \wedge du^k$ on the manifold M. If we define the coefficients $b_{ijk}(u)$ by formula (3.42), then, as shown by straightforward calculations, all relations (3.25), (3.26), (3.27), and (3.29) are satisfied.

It follows from formula (3.17) that a first-order homogeneous 2-form (3.21) is exact if and only if the following relations are valid:

$$g_{ij} = f_{ij} + f_{ji}, \qquad (3.43)$$

$$b_{ijk} = \frac{\partial f_{ji}}{\partial u^k} + \frac{\partial f_{ik}}{\partial u^j} - \frac{\partial f_{jk}}{\partial u^i}, \qquad (3.44)$$

where $f_{ij}(u)$ is an arbitrary smooth covariant tensor field of rank 2 on the manifold M.

Let us introduce a new skew-symmetric tensor field $\beta_{ij}(u) = f_{ji}(u) - f_{ij}(u)$. Then we find that a first-order homogeneous 2-form (3.21) is exact if and only if $g_{ij}(u)$ is an arbitrary smooth symmetric tensor field on M and the coefficients $b_{ijk}(u)$ can be represented in the form

$$b_{ijk} = \frac{1}{2}\left(\frac{\partial g_{ji}}{\partial u^k} + \frac{\partial g_{ik}}{\partial u^j} - \frac{\partial g_{jk}}{\partial u^i} - \frac{\partial \beta_{ji}}{\partial u^k} - \frac{\partial \beta_{ik}}{\partial u^j} + \frac{\partial \beta_{jk}}{\partial u^i}\right), \quad (3.45)$$

where $\beta_{ij}(u)$ is an arbitrary smooth skew-symmetric tensor field on the manifold M.

Thus,

$$H^2_{[1]}(\Omega M, \mathbf{R}) = H^3(M, \mathbf{R}),$$

and Theorems 3.2 and 3.1 are thereby proved.

We note that homogeneous k-forms of order 1 are invariant with respect to the action of the group $\mathrm{Diff}^+(S^1)$ of diffeomorphisms of the circle S^1 preserving the orientation. This is important, in particular, from the viewpoint of applications in the theory of closed boson strings in curved N-dimensional space-time M with the metric (gravitational field) g_{ij}. The configuration space of these closed boson strings is the loop space ΩM of the pseudo-Riemannian manifold (M, g_{ij}) (the requirement of invariance or, in other words, independence on the parametrizations of the loops is necessary in the theory of boson strings for physically sensible objects on the configuration space ΩM).

The homogeneous (pre)-symplectic structures on loop spaces of smooth manifolds studied by the present author in [109], [110], [115], [124], [126] (see Sections 2.2, 2.4) correspond exactly to the closed homogeneous 2-forms on ΩM, that is, they are elements of $Z^2_{[m]}(\Omega M)$.

The approach developed in this section gives the possibility of finding explicitly (of expressing via the de Rham cohomology of the manifold) the higher ($i > 2$) homogeneous cohomology groups $H^i_{[m]}(\Omega M, \mathbf{R})$ of the loop space, including the cases of higher orders $m \geq 2$, although this requires huge calculations.

We note that, generally speaking, the shift equal to m between the homogeneous cohomology groups of loop spaces of manifolds

$H^i_{[m]}(\Omega M, \mathbf{R})$ and the de Rham cohomology groups of the manifolds $H^i(M, \mathbf{R})$ in formulae (3.7) ($m = 0$) and (3.9) ($m = 1$) is not a general property for any i and m in the theory of homogeneous cohomology. In any case, it is easy to show that

$$H^0_{[2]}(\Omega M, \mathbf{R}) = 0.$$

Indeed, homogeneous 0-forms of the second order are functionals of the form

$$F = \int_{S^1} b_{ij}(u) u^i_x u^j_x dx, \qquad (3.46)$$

where $b_{ij}(u)$ is an arbitrary smooth symmetric tensor field on the manifold M.

Let us find the conditions for the second-order homogeneous 0-form (3.46) to be closed:

$$(dF)(\xi) = \int_{S^1} \xi^i \left(\frac{\partial b_{kj}}{\partial u^i} u^k_x u^j_x - (b_{ij} u^j_x)_x - (b_{ji} u^j_x)_x \right) dx =$$

$$= \int_{S^1} \xi^i \left[(-b_{ij} - b_{ji}) u^j_{xx} + \right.$$

$$\left. + \left(\frac{\partial b_{kj}}{\partial u^i} - \frac{\partial b_{ij}}{\partial u^k} - \frac{\partial b_{ji}}{\partial u^k} \right) u^k_x u^j_x \right] dx \equiv 0.$$

Thus, for any closed second-order homogeneous 0-form (3.46), the relation $b_{ij} = 0$ must be satisfied, that is, any closed second-order homogeneous 0-form must be equal to 0 identically.

4 Local and non-local Poisson structures of differential-geometric type

4.1 Riemannian geometry of multidimensional local Poisson structures of hydrodynamic type

4.1.1 Multidimensional local Poisson brackets of hydrodynamic type

Let us consider *multidimensional systems of hydrodynamic type*, in other words, multidimensional evolution quasilinear systems of first-order partial differential equations:

$$u_t^i = v_j^{i\alpha}(u) \frac{\partial u^j}{\partial x^\alpha}\,, \tag{4.1}$$

where $i, j = 1, \ldots, N$; $\alpha = 1, \ldots, n$; $x = (x^1, \ldots, x^n)$; here for all α the coefficients $v_j^{i\alpha}(u)$ are arbitrary $N \times N$ matrix functions depending on $u = (u^1, \ldots, u^N)$, $u^i = u^i(x, t)$.

Such systems naturally arise not only in Euler hydrodynamics and gas dynamics but also in the theory of N-layer flows (the Benney equations), as dispersionless limits of different physical systems, as the result of averaging by the Whitham method, and so on. As a rule, systems of hydrodynamic type arising in physically interesting conservative cases are Hamiltonian. A general differential-geometric Hamiltonian approach to systems of the form (4.1) was proposed by Dubrovin and Novikov [34], [35].

Multidimensional Hamiltonian systems of hydrodynamic type introduced in [34], [35] have the form

$$u_t^i = \{u^i(x), H\}, \tag{4.2}$$

where H is a *functional of hydrodynamic type*, that is, its density $h(u(x))$ is a function that depends only on the field variables $u^i(x)$ and does not depend on their derivatives:

$$H = \int h(u(x))\, d^n x, \tag{4.3}$$

and the *multidimensional local homogeneous Poisson bracket of hydrodynamic type* has the form

$$\{u^i(x), u^j(y)\} = g^{ij\alpha}(u(x))\delta_\alpha(x-y) +$$
$$+u^k_\alpha(x)b^{ij\alpha}_k(u(x))\delta(x-y), \qquad (4.4)$$

where $i, j, k = 1, \ldots, N$; $\alpha = 1, \ldots, n$; $x = (x^1, \ldots, x^n)$;

$$u^i_\alpha = \frac{\partial u^i}{\partial x^\alpha}.$$

The corresponding multidimensional Hamiltonian operator is a homogeneous multidimensional differential operator of the first order:

$$K^{ij}[u(x)] = g^{ij\alpha}(u(x))\frac{\partial}{\partial x^\alpha} + u^k_\alpha b^{ij\alpha}_k(u(x)),$$

and the Hamiltonian system (4.2) can be rewritten in the form

$$u^i_t = K^{ij}[u(x)]\frac{\delta H}{\delta u^j(x)}.$$

It is clear that if a Poisson bracket has the form (4.4), then for any functional of hydrodynamic type (4.3) the corresponding Hamiltonian system (4.2) is a multidimensional evolution quasilinear system of first-order partial differential equations, that is, of the form (4.1). Moreover, the forms of a Poisson bracket (4.4) and a system of hydrodynamic type (4.1) are preserved under local changes $u^i = u^i(v)$ of coordinates on a manifold M, and the coefficients of the bracket and the matrices of the system are transformed as differential-geometric objects on M.

The matrices $v^{i\alpha}_j(u)$ of the system (4.1) are transformed as *affinors*, that is, tensors of the type $(1, 1)$. The Hamiltonian operator is transformed as follows:

$$K^{ij}[u(v)] = \frac{\partial u^i}{\partial v^s}\widetilde{K}^{sr}[v]\frac{\partial u^j}{\partial v^r}.$$

Correspondingly, for the coefficients of the Hamiltonian operator we obtain the following transformation formulae:

$$g^{ij\alpha}(u(v)) = \frac{\partial u^i}{\partial v^s}\tilde{g}^{sr\alpha}\frac{\partial u^j}{\partial v^r},$$

$$b_q^{ij\alpha}(u(v)) = \frac{\partial u^i}{\partial v^s} \frac{\partial^2 u^j}{\partial v^r \partial v^p} \frac{\partial v^p}{\partial u^q} \tilde{g}^{sr\alpha}(v) + \frac{\partial u^i}{\partial v^s} \frac{\partial u^j}{\partial v^r} \frac{\partial v^p}{\partial u^q} \tilde{b}_p^{sr\alpha}(v).$$

This makes it possible to develop a general differential-geometric approach to Hamiltonian systems of hydrodynamic type (4.1).

If $\det(g^{ij\alpha}) \neq 0$ for all α, then multidimensional Poisson structures of hydrodynamic type (4.4) are called *non-degenerate*.

The coefficients $g^{ij\alpha}(u)$ are transformed as metrics with upper indices on the manifold. For non-degenerate Poisson brackets of the form (4.4) we introduce the coefficients $\Gamma_{sk}^{j\alpha}(u)$ by the formulae

$$b_k^{ij\alpha}(u) = -g^{is\alpha}(u)\Gamma_{sk}^{j\alpha}(u).$$

The coefficients $\Gamma_{sk}^{j\alpha}(u)$ are transformed like the Christoffel symbols of an affine connection on the manifold:

$$\Gamma_{pq}^{j\alpha}(u(v)) = \frac{\partial u^j}{\partial v^k}\left(\tilde{\Gamma}_{rs}^{k\alpha}(v)\frac{\partial v^s}{\partial u^q}\frac{\partial v^r}{\partial u^p} + \frac{\partial^2 v^k}{\partial u^p \partial u^q}\right).$$

It follows from the results of Dubrovin and Novikov [34], [35] that for non-degenerate local Poisson brackets of the form (4.4) all the metrics $g^{ij\alpha}(u)$ are flat metrics on the manifold M and all the coefficients $\Gamma_{ik}^{j\alpha}(u)$ define symmetric affine connections of zero Riemannian curvature, which are compatible with the metrics $g^{ij\alpha}(u)$ corresponding to them. In fact, in the one-dimensional case (for $n = 1$) Dubrovin and Novikov proved the following theorem (we omit the index α of all the coefficients of the Poisson bracket in the one-dimensional case).

Theorem 4.1 (Dubrovin and Novikov [34]) *If* $\det[g^{ij}(u)] \neq 0$ *and* $n = 1$, *then the expression (4.4) defines a Poisson bracket if and only if*

1. $g^{ij}(u)$ *is a Riemannian or pseudo-Riemannian metric of zero curvature (that is, it is simply a flat metric),*

2. $b_k^{ij}(u) = -g^{is}(u)\Gamma_{sk}^j(u)$, *where* $\Gamma_{sk}^j(u)$ *are the coefficients of the Riemannian connection generated by the metric* $g^{ij}(u)$, *that is, the symmetric connection compatible with the metric (the Levi–Civita connection).*

Thus, for $n = 1$ there always exist local coordinates $v^i = v^i(u)$ in which the one-dimensional Poisson bracket of the form (4.4) is constant and has the form:

$$\{v^i(x), v^j(y)\} = \varepsilon^i \delta^{ij} \delta_x(x - y),$$

where $\varepsilon^i = \pm 1$, $i = 1, \ldots, N$.

It is easy to show that each multidimensional Poisson bracket of hydrodynamic type (4.4) is always a sum over α of corresponding one-dimensional Poisson brackets (see the right-hand side of formula (4.4) for each α separately). Moreover, all these one-dimensional Poisson brackets will be compatible if for each α we put $x^\alpha = x$. In fact, there is always a reduction $x^\alpha = c^\alpha x$, where c^α are arbitrary non-zero constants. This reduction preserves skew-symmetry and the Jacobi identity. As a result of this reduction, any multidimensional Poisson bracket (4.4) is transformed to an arbitrary linear combination of the corresponding one-dimensional Poisson brackets.

Generally speaking, the condition that an expression of the form (4.4) is a multidimensional Poisson bracket is a considerably stronger condition for the coefficients than the necessary requirement that all the corresponding one-dimensional Poisson brackets are compatible.

Thus, the problem of describing or classifying multidimensional homogeneous Poisson brackets of hydrodynamic type corresponds to the problem of describing or classifying an important and interesting special subclass of compatible one-dimensional homogeneous Poisson brackets of hydrodynamic type. The general problem of describing all compatible one-dimensional homogeneous Poisson brackets of hydrodynamic type was recently solved by the present author in [133]–[137] (see also [29], [118], [119], [128]–[140] on this interesting problem). This problem is very interesting and topical in the theory of one-dimensional systems of hydrodynamic type and their applications.

The problem of describing of multidimensional Poisson brackets of the form (4.4) has not yet been solved completely, but for the case of a small number of components $N \leq 4$ the present author has obtained exhaustive results including a complete algebraic classification. In particular, a complete explicit description and classification of two-dimensional homogeneous Poisson brackets of hydrodynamic type for a small number of components $1 \leq N \leq 4$ was obtained by the present author in [105]. It is interesting that for $N = 4$ an infinite one-parameter family of canonical two-dimensional Poisson structures of the form (4.4), which cannot be reduced to each other by local changes of coordinates on the manifold, arises for the first time (for $1 \leq N < 4$ there are always only finitely many *canonical* Poisson structures, and all the others can be reduced to them by local changes of coordinates on the manifold).

First of all, let us find general relations for the coefficients of the bracket (4.4).

Lemma 4.1 *The expression (4.4) defines a Poisson bracket if and only if the following relations are fulfilled:*

$$g^{ij\alpha} = g^{ji\alpha}, \tag{4.5}$$

$$\frac{\partial g^{ij\alpha}}{\partial u^k} = b_k^{ij\alpha} + b_k^{ji\alpha}, \tag{4.6}$$

$$\sum_{(\alpha,\beta)} (g^{si\alpha} b_s^{jr\beta} - g^{sj\beta} b_s^{ir\alpha}) = 0, \tag{4.7}$$

$$\sum_{(i,j,r)} (g^{si\alpha} b_s^{jr\beta} - g^{sj\beta} b_s^{ir\alpha}) = 0, \tag{4.8}$$

$$\sum_{(\alpha,\beta)} \left\{ g^{si\alpha} \left(\frac{\partial b_s^{jr\beta}}{\partial u^q} - \frac{\partial b_q^{jr\beta}}{\partial u^s} \right) + b_s^{ij\alpha} b_q^{sr\beta} - b_s^{ir\alpha} b_q^{sj\beta} \right\} = 0, \tag{4.9}$$

$$g^{si\beta} \frac{\partial b_q^{jr\alpha}}{\partial u^s} - b_s^{ij\beta} b_q^{sr\alpha} - b_s^{ir\beta} b_q^{js\alpha} =$$
$$= g^{sj\alpha} \frac{\partial b_q^{ir\beta}}{\partial u^s} - b_s^{ji\alpha} b_q^{sr\beta} - b_q^{is\beta} b_s^{jr\alpha}, \tag{4.10}$$

$$\frac{\partial}{\partial u^k} \left\{ g^{si\alpha} \left(\frac{\partial b_s^{jr\beta}}{\partial u^q} - \frac{\partial b_q^{jr\beta}}{\partial u^s} \right) + b_s^{ij\alpha} b_q^{sr\beta} - b_s^{ir\alpha} b_q^{sj\beta} \right\} +$$
$$+ \sum_{(i,j,r)} \left\{ b_q^{si\beta} \left(\frac{\partial b_k^{jr\alpha}}{\partial u^s} - \frac{\partial b_s^{jr\alpha}}{\partial u^k} \right) \right\} +$$

$$+\frac{\partial}{\partial u^q}\left\{g^{si\beta}\left(\frac{\partial b_s^{jr\alpha}}{\partial u^k}-\frac{\partial b_k^{jr\alpha}}{\partial u^s}\right)+b_s^{ij\beta}b_k^{sr\alpha}-b_s^{ir\beta}b_k^{sj\alpha}\right\}+$$

$$+\sum_{(i,j,r)}\left\{b_k^{si\alpha}\left(\frac{\partial b_q^{jr\beta}}{\partial u^s}-\frac{\partial b_s^{jr\beta}}{\partial u^q}\right)\right\}=0. \qquad (4.11)$$

Relations (4.5) and (4.6) are equivalent to skew-symmetry of the bracket (4.4), and relations (4.7)–(4.11) are equivalent to the Jacobi identity for a skew-symmetric bracket of the form (4.4).

In Lemma 4.1 the non-degeneracy of the bracket is not assumed and no additional conditions are imposed on the coefficients of the bracket (4.4). All relations (4.5)–(4.11) are obtained from the condition of skew-symmetry of the bracket (4.4) and the fulfilment of the Jacobi identity for this bracket on arbitrary linear functionals of the form $I = \int f_i(x)u^i(x)\,d^n x$ by direct calculations.

In particular, it follows immediately from Lemma 4.1 that each multidimensional Poisson bracket (4.4) is always a sum of one-dimensional Poisson brackets corresponding to each independent variable x^α. Moreover, using Lemma 4.1 it is easy to prove the compatibility of the corresponding one-dimensional Poisson brackets given by the multidimensional Poisson bracket of hydrodynamic type.

First of all, let us consider relations (4.5)–(4.11) in the one-dimensional case $(n = 1)$. It is clear that in the one-dimensional case relation (4.8) is a consequence of relation (4.7) and, furthermore, relation (4.10) follows from relations (4.9), (4.7), and (4.6).

For a non-degenerate metric $g^{ij\alpha}$ we have: condition (4.5) gives symmetry of the metric; condition (4.6) means that the connection $\Gamma_{ij}^{k\alpha} = -g_{is}^\alpha b_j^{sk\alpha}$ is compatible with the metric $g^{ij\alpha}$, that is, the covariant derivative of the metric is equal to zero; condition (4.7) is equivalent to the condition that the connection $\Gamma_{ij}^{k\alpha}$ is symmetric, that is, $\Gamma_{ij}^{k\alpha} = \Gamma_{ji}^{k\alpha}$; condition (4.9) means precisely that the connection is flat, that is, the Riemannian curvature tensor of the connection is equal to zero; condition (4.11) for $\alpha = \beta$ and for a non-degenerate metric $g^{ij\alpha}$ follows from relations (4.7) and (4.9). This proves the Dubrovin–Novikov theorem for non-degenerate one-dimensional local Poisson structures of hydrodynamic type.

4.1.2   Tensor obstructions for multidimensional local Poisson brackets of hydrodynamic type

The following theorem gives a complete set of differential-geometric relations for the coefficients of the bracket (4.4) such that expression (4.4) defines a non-degenerate Poisson bracket if and only if these relations are satisfied.

Theorem 4.2 *Flat non-degenerate metrics $g^{ij\alpha}(u)$ and the Riemannian connections $\Gamma^{i\alpha}_{jk}$ defined by these metrics (the Levi–Civita connections) generate a multidimensional Poisson structure (4.4) if and only if the tensors $T^{i\alpha\beta}_{jk} = \Gamma^{i\beta}_{jk} - \Gamma^{i\alpha}_{jk}$ satisfy the following relations:*

$$T^{ijk\alpha\beta} = T^{kji\alpha\beta}, \qquad (4.12)$$

where $T^{ijk\alpha\beta} = -g^{ks\beta}T^{ij\alpha\beta}_s = -g^{ks\beta}g^{ip\alpha}T^{j\alpha\beta}_{ps}$, $T^{ij\alpha\beta}_k = g^{is\alpha}T^{j\alpha\beta}_{sk}$,

$$\sum_{(i,j,k)} T^{ijk\alpha\beta} = 0, \qquad (4.13)$$

where $\sum_{(i,j,k)}$ means summation over all permutations of the elements (i,j,k),

$$T^{ijs\alpha\beta}T^{r\alpha\beta}_{st} = T^{irs\alpha\beta}T^{j\alpha\beta}_{st}, \qquad (4.14)$$

$$\nabla^\alpha_t T^{ijk\alpha\beta} = 0, \qquad (4.15)$$

where ∇^α_t is the corresponding covariant derivative (note that in all these relations there is no summation over α and β).

The tensors $T^{ijk\alpha\beta}(u)$ are obstructions to reducing the non-degenerate Poisson structures (4.4) to a constant form, that is, a non-degenerate Poisson bracket of the form (4.4) can be reduced to a constant bracket by a local change of coordinates if and only if all the tensors $T^{ijk\alpha\beta}(u)$ are equal to zero identically. In fact, if even one of these tensors is not equal to zero identically, then the Poisson bracket cannot be reduced to a constant form by local change of coordinates, since for each constant bracket $\Gamma^{i\alpha}_{jk}(u) = 0$ and therefore in these local coordinates all the tensors $T^{i\alpha\beta}_{jk}(u)$ are equal to zero identically. The converse is also obvious: if all the tensors $T^{i\alpha\beta}_{jk}(u)$ are equal to zero identically, then all the connections $\Gamma^{i\alpha}_{jk}(u)$ are equal to each other, and consequently all of them are equal to zero in the flat coordinates of the metric $g^{ij1}(u)$, and all the metrics are constant in these coordinates by virtue of their compatibility with the corresponding connections.

Relations (4.12) and (4.14) were found by Dubrovin and Novikov in [35], where they described all tensor relations arising for the coefficients of Poisson structures of hydrodynamic type (4.4) from the fulfilment of the Jacobi identity for the bracket on all functionals of hydrodynamic type (4.3). In the present author's paper [105] it is proved that this set of tensor relations is not complete, and all necessary and sufficient conditions for the expression (4.4) to be a non-degenerate Poisson structure were found. Theorem 4.2 describes the complete set of tensor relations for the coefficients of a non-degenerate Poisson bracket of the form (4.4) that guarantee skew-symmetry of the bracket and the fulfilment of the Jacobi identity for arbitrary functionals.

As is shown below, relations (4.13) and (4.15) are essential and play an important role in the classification of multidimensional Poisson brackets of hydrodynamic type. It is curious that in the one-dimensional case, as Dubrovin and Novikov showed in [34], the relations obtained for a non-degenerate skew-symmetric one-dimensional bracket (4.4) from the fulfilment of the Jacobi identity for the bracket only for functionals of hydrodynamic type guarantee the fact that the one-dimensional bracket (4.4) can be reduced to a constant form by a local change of coordinates, and consequently defines a Poisson bracket. As Theorem 4.2 shows, even in the two-dimensional case not all necessary relations can be obtained on functionals of hydrodynamic type.

If all the metrics $g^{ij\alpha}$ are non-degenerate, then from Lemma 4.1 we deduce that for $\alpha \neq \beta$ condition (4.7) gives relation (4.12) for the obstruction tensors; condition (4.8) is equivalent to relation (4.13); condition (4.9) is equivalent to relation (4.14); condition (4.10) is equivalent to relation (4.15); for the case of non-degenerate metrics condition (4.11) is a direct consequence of relations (4.5)–(4.10) (in the general case, this is not true, that is, condition (4.11) is essential for the case of degenerate metrics).

For $N = 1$ and any n all the obstruction tensors are equal to zero identically. In fact, in the one-component case relations (4.5), (4.7), (4.9), and (4.11) are automatically fulfilled, relation (4.6) gives $\partial g^\alpha / \partial u = 2b^\alpha$, and from the relation (4.8) we deduce that $g^\alpha b^\beta = g^\beta b^\alpha$, and relation (4.10) follows from (4.6) and (4.8). Thus, in the one-component case for any n and for any α we have $g^\alpha(u) = c^\alpha g(u)$, where $g^\alpha(u)$ is an arbitrary function, and c^α is an arbitrary constant. Correspondingly, all the obstruction tensors $T^{\alpha\beta}$ are equal to zero identically and therefore all multidimensional one-component Poisson brackets of the form (4.4) can be reduced to a constant form by a local change of the unique coordinate u^1.

4.1.3   Infinite-dimensional Lie algebras associated with multidimensional local Poisson brackets of hydrodynamic type

A flat metric $g^{ij\alpha}(u)$ can always be reduced to a constant form by a local change of the variables u^i. Let us reduce one of the metrics (for definiteness, we reduce the first one) to a constant form. The following important theorem is a simple consequence of Theorem 4.2.

Theorem 4.3 ([35], [105]) *If $g^{ij1} = $ const for a non-degenerate multidimensional Poisson structure (4.4), then all the other metrics of the bracket are linear with respect to the fields $u^i(x)$:*

$$g^{ij\alpha} = (c_k^{ij\alpha} + c_k^{ji\alpha})u^k + g_0^{ij\alpha}, \qquad c_k^{ij\alpha} = \text{const}, \quad g_0^{ij\alpha} = \text{const}.$$

Proof. Let the metric g^{ij1} be constant. Then we have $\Gamma_{jk}^{i1} = 0$ in these local coordinates. It follows from relation (4.15) that $T^{ijk1\alpha} = const$ for all α in these local coordinates.

Let us prove that all the coefficients $b_k^{ij\alpha}$ are also constant in these local coordinates. In fact,

$$b_k^{ij\alpha} = -g^{is\alpha}\Gamma_{sk}^{j\alpha} = -g^{is\alpha}T_{sk}^{j1\alpha} =$$
$$= -g_{kq1}g^{is\alpha}g^{qr1}T_{rs}^{j1\alpha} = g_{kq1}T^{qji1\alpha} = const.$$

Here, we use that all affine connections $\Gamma_{jk}^{i\alpha}$ are symmetric.

Now the linearity of all the metrics of the bracket in these local coordinates follows from relation (4.6). $\quad\blacksquare$

Remark 4.1 For $N \geq 3$ Theorem 4.3 was proved by Dubrovin and Novikov in [35]. The study of the special cases $N = 1, 2$ in [35] is not correct, since the study is based, as is mentioned above, on the set of relations for the obstruction tensors, which is obtained in [35] and which is incomplete and insufficient in order to guarantee that the expression (4.4) is a Poisson bracket. In [35] there are found only those relations that follow from the fact that the Jacobi identity is fulfilled for a non-degenerate bracket of the form (4.4) for all functionals of hydrodynamic type. This is not sufficient for the fulfilment of the Jacobi identity for all functionals. But it is very curious that in the multicomponent case, that is, for $N \geq 3$, this incomplete set of tensor relations is nevertheless sufficient to prove the fact that any non-degenerate Poisson bracket of the form (4.4) can be reduced to a linear form (Theorem 4.3) (but, of course, this set of tensor relations is not sufficient for the classification of such Poisson structures or corresponding infinite-dimensional Lie algebras).

Corollary 4.1 *An arbitrary non-degenerate multidimensional local Poisson structure of hydrodynamic type (4.4) is defined by an infinite-dimensional Lie algebra of special type and a 2-cocycle of special type on this Lie algebra:*

$$[\xi, \eta]_k = c_k^{ij\alpha}((\eta_i)_\alpha \xi_j - \eta_j(\xi_i)_\alpha),$$

$$\xi = (\xi_1, \ldots, \xi_N), \quad \xi_i(x) \in C^1(T^n).$$

The corresponding 2-cocycles on the Lie algebra must have the following special form:

$$\langle \xi, \eta \rangle = \int g_0^{ij\alpha}(\eta_j(x))_\alpha \xi_i(x) \, d^n x.$$

Let us mention briefly here a general scheme that goes back to Sophus Lie (see [69]) concerning interconnections between Lie algebras and Poisson structures whose coefficients depend linearly (possibly, non-homogeneously) on coordinates (the Lie–Poisson brackets). In the general infinite-dimensional case we describe special infinite-dimensional Lie algebras corresponding to arbitrary Poisson structures whose coefficients depend linearly (possibly, non-homogeneously) on the field variables $u^i(x)$ and their derivatives $u^i_{(k)}$, $k = (k_1, \ldots, k_n)$, where $u^i_{(k)} = \partial^{|k|} u^i / \partial (x^1)^{k_1} \cdots \partial (x^n)^{k_n}$, $|k| = k_1 + \cdots + k_n$. In what follows we shall often use this scheme in different situations.

Let us consider Hamiltonian operators of the form

$$M^{ij} = (a_s^{ij,(k)(p)} u_{(k)}^s + b^{ij,(p)}) \frac{\partial^{|p|}}{\partial(x^1)^{p_1} \cdots \partial(x^n)^{p_n}}, \qquad (4.16)$$

where $a_s^{ij,(k)(p)}$ and $b^{ij,(p)}$ are constants. Consider the space S of sequences $(\xi_1, \ldots, \xi_N)$, $\xi_i \in C^\infty(T^n)$. If ξ and η belong to S, then we have

$$(\xi, M(\eta)) \equiv \int_{T^n} \xi_i M^{ij} \eta_j \, d^n x =$$

$$= \int_{T^n} u^s [\xi, \eta]_s \, d^n x + \int_{T^n} \xi_i b^{ij,(p)} \eta_{j(p)} \, d^n x. \qquad (4.17)$$

Correspondingly, on the space S a bilinear operation $[\,\cdot\,,\cdot\,]$:

$$(\xi, \eta) \mapsto \zeta = [\xi, \eta] \in S, \qquad (4.18)$$

and a bilinear form

$$\omega(\xi, \eta) = \int_{T^n} \xi_i b^{ij,(p)} \eta_{j(p)} \, d^n x \qquad (4.19)$$

are defined.

The operator M^{ij} of the form (4.16) is skew-symmetric if and only if the bilinear operation (4.18) and the bilinear form (4.19) are skew-symmetric on the space S, that is,

$$[\xi, \eta] = -[\eta, \xi]$$

and

$$\omega(\xi, \eta) = -\omega(\eta, \xi).$$

The operator M^{ij} of the form (4.16) is Hamiltonian if and only if the space S is a Lie algebra with respect to the bilinear operation (4.18), that is, this operation is skew-symmetric and satisfies the Jacobi identity

$$[\xi, [\eta, \zeta]] + [\eta, [\zeta, \xi]] + [\zeta, [\xi, \eta]] = 0,$$

and, in addition, the bilinear form (4.19) is a 2-cocycle on this Lie algebra, that is, this bilinear form is skew-symmetric and satisfies the closedness identity

$$(d\omega)(\xi, \eta, \zeta) \equiv \omega([\xi, \eta], \zeta) + \omega([\eta, \zeta], \xi) + \omega([\zeta, \xi], \eta) = 0. \qquad (4.20)$$

For any k-form $\omega(a_1, \ldots, a_k)$ on a Lie algebra $\mathcal{G}$, $a_i \in \mathcal{G}$, the differential d is defined by

$$(d\omega)(a_1, \ldots, a_{k+1}) = \qquad\qquad\qquad (4.21)$$
$$= \sum_{i<j} (-1)^{i+j+1} \omega([a_i, a_j], a_1, \ldots, \hat{a}_i, \ldots, \hat{a}_j, \ldots, a_{k+1}),$$

with $d^2\omega = 0$.

A 2-cocycle $\omega(\xi, \eta)$ defined by a Hamiltonian operator of the form (4.16) is *cohomologous to zero*, that is,

$$\omega(\xi, \eta) = (df)(\xi, \eta) \equiv f([\xi, \eta]),$$

where f is a 1-form on the Lie algebra S, if and only if it can be annihilated by a shift of the field variables $u^i \mapsto u^i - c^i$, where c^i are arbitrary constants, that is, provided that

$$b^{ij,(p)} = a_k^{ij,(0)(p)} c^k. \qquad\qquad\qquad (4.22)$$

Example 4.1 *Poisson structures defined by the Lie algebra of vector fields on the n-dimensional torus T^n.*

The commutator of vector fields ξ and η has the form:

$$[\xi, \eta]_k = \xi_s \frac{\partial \eta_k}{\partial x^s} - \eta_s \frac{\partial \xi_k}{\partial x^s}. \qquad\qquad\qquad (4.23)$$

For the corresponding Hamiltonian operator M^{ij} we obtain:

$$\int_{T^n} \xi_i M^{ij} \eta_j \, d^n x = \int_{T^n} u^k [\xi, \eta]_k \, d^n x =$$
$$= \int_{T^n} \xi_i \left(u^i \frac{\partial}{\partial x^j} + u^j \frac{\partial}{\partial x^i} + \frac{\partial u^i}{\partial x^j} \right) \eta_j \, d^n x, \qquad (4.24)$$

$$M^{ij} = u^i \frac{\partial}{\partial x^j} + u^j \frac{\partial}{\partial x^i} + \frac{\partial u^i}{\partial x^j}. \qquad\qquad\qquad (4.25)$$

It follows from the relations of Lemma 4.1 that if for a 2-cocycle on the Lie algebra of vector fields on T^n (4.23) the corresponding Poisson structure remains in the class of local multidimensional homogeneous Poisson structures of hydrodynamic type (4.4) (see formulae (4.17) and (4.19)), then this 2-cocycle is cohomologous to zero. For $n \leq 2$ the Poisson structure (4.25) is non-degenerate, but for $n > 2$ all the metrics in (4.25) are degenerate.

Let us consider now the first non-trivial case of multidimensional Poisson structures of the form (4.4), namely, the two-component case $N = 2$.

Theorem 4.4 ([105]) *If for $N = n = 2$ for a non-degenerate Poisson structure (4.4) the obstruction tensor $T_{jk}^{i12}(u)$ is not equal to zero identically, that is, this Poisson structure cannot be reduced to a constant form by a local change of coordinates, then by a local change of coordinates the Poisson structure can be reduced to the canonical form generated by flat metrics*

$$g^{ij1} = \begin{pmatrix} 1 & 0 \\ 0 & -1 \end{pmatrix}, \qquad g^{ij2} = \begin{pmatrix} 2v & u+v \\ u+v & 2u \end{pmatrix}. \qquad (4.26)$$

In particular, if one of the metrics of a local two-dimensional two-component Poisson bracket of hydrodynamic type is positive (or negative) definite, then this Poisson structure can be reduced to a constant form. The Poisson structure generated by the canonical flat metrics (4.26) is connected with the Lie algebra (4.23) of vector fields on the two-dimensional torus T^2.

Proof. Let us consider canonical coordinates w^i, in which the metric g^{ij1} is constant. According to Theorem 4.3 the metric g^{ij2} must have a linear form in these canonical coordinates:

$$g^{ij2} = (b_k^{ij2} + b_k^{ji2})w^k + c^{ij} + c^{ji}, \quad b_k^{ij2} = const, \quad c^{ij} = const.$$

The condition that the affine connection Γ_{jk}^{i2} defined by the metric g^{ij2} is symmetric and has zero Riemannian curvature gives the relations

$$g^{si2}b_s^{jr2} = g^{sj2}b_s^{ir2}, \qquad (4.27)$$

$$b_s^{ij2}b_k^{sr2} = b_s^{ir2}b_k^{sj2}. \qquad (4.28)$$

For the components of the obstruction tensor T_{jk}^{i12} in the canonical coordinates w^i we obtain the relations:

$$T_{jk}^{i12} = \Gamma_{jk}^{i2}, \quad T^{ijk12} = g^{ip1}b_p^{kj2}, \quad b_k^{ij2} = -g^{is2}\Gamma_{sk}^{j2}, \quad \Gamma_{jk}^{i1} = 0.$$

Relations (4.12)–(4.15) for the obstruction tensor give the following relations:

$$g^{ip1}b_p^{kj2} = g^{kp1}b_p^{ij2}, \qquad (4.29)$$

$$g^{ip1}b_p^{kj2} + g^{jp1}b_p^{ik2} + g^{kp1}b_p^{ji2} = 0. \qquad (4.30)$$

The identity (4.14)

$$T^{ijs12}T_{sk}^{r12} = T^{irs12}T_{sk}^{j12} \qquad (4.31)$$

follows from relation (4.28) in this case. We obtain from (4.28)

$$b_p^{sj2}\Gamma_{ks}^{r2} = b_p^{sr2}\Gamma_{ks}^{j2},$$

and relation (4.31) is equivalent to the relation

$$g^{ip1}b_p^{sj2}\Gamma_{sk}^{r2} = g^{ip1}b_p^{sr2}\Gamma_{sk}^{j2}$$

or

$$b_p^{sj2}\Gamma_{sk}^{r2} = b_p^{sr2}\Gamma_{sk}^{j2}.$$

Proving Theorem 4.3 we also proved that $T^{ijk12} = const$ in the canonical coordinates w^i.

Relation (4.15) follows from the relations (4.27)–(4.30):

$$\nabla_p^1 T^{ijk12} = 0,$$

$$\nabla_p^2 T^{ijk12} = \Gamma_{sp}^{i2}T^{sjk12} + \Gamma_{sp}^{j2}T^{isk12} + \Gamma_{sp}^{k2}T^{ijs12} =$$
$$= T_{sp}^{i12}T^{sjk12} + T_{sp}^{j12}T^{isk12} + T_{sp}^{k12}T^{ijs12} =$$
$$= T_{sp}^{j12}(T^{kis12} + T^{isk12} + T^{ski12}) = 0.$$

Let us consider two possible cases separately.

1) The metric g^{ij1} is positive or negative definite (with the signature $(+,+)$ or $(-,-)$). In this case the metric can be reduced to the form

$$(g^{ij1}) = \pm(\delta^{ij})$$

by local changes of coordinates.

2) The metric g^{ij1} is indefinite with the signature $(+,-)$. The metric can be reduced to the form

$$(g^{ij}) = \begin{pmatrix} 1 & 0 \\ 0 & -1 \end{pmatrix}$$

by local changes of coordinates.

Let us prove that in the first case $b_k^{ij\,2} = 0$, the obstruction tensor $T_{jk}^{i\,12}$ is equal to zero identically, and the Poisson structure under consideration is constant in the canonical coordinates w^i.

Relation (4.29) in this case gives

$$b_i^{kj\,2} = b_k^{ij\,2},$$

and from relation (4.30) we obtain

$$b_i^{kj\,2} + b_j^{ik\,2} + b_k^{ji\,2} = 0,$$

and, consequently,

$$b_1^{11,2} = b_2^{22,2} = 0, \qquad b_2^{11,2} + b_1^{21,2} + b_1^{12,2} = 0,$$

$$b_1^{22,2} + b_2^{12,2} + b_2^{21,2} = 0, \qquad b_1^{21,2} = b_2^{11,2},$$

$$b_2^{12,2} = b_1^{22,2}, \qquad b_1^{12,2} = -2b_2^{11,2}, \qquad b_2^{21,2} = -2b_1^{22,2}.$$

From identity (4.28) for $i = 1$, $j = 1$, $k = 2$, $r = 2$ we obtain

$$b_2^{11,2} b_1^{12,2} + b_2^{21,2} b_2^{12,2} = 0$$

or

$$(b_2^{11,2})^2 + (b_1^{22,2})^2 = 0.$$

Thus, in this case we have $b_k^{ij,2} = 0$.

Let us consider now the indefinite case 2):

$$g^{ij\,1} = (-1)^{j+1} \delta^{ij}.$$

Relation (4.29) gives

$$(-1)^{i+1} b_i^{kj\,2} = (-1)^{k+1} b_k^{ij\,2},$$

that is,

$$b_1^{2j\,2} = -b_2^{1j\,2}.$$

From relation (4.30) we obtain

$$(-1)^{i+1} b_i^{kj\,2} + (-1)^{j+1} b_j^{ik\,2} + (-1)^{k+1} b_k^{ji\,2} = 0,$$

that is,

$$b_1^{11,2} = b_2^{22,2} = 0, \quad b_2^{21,2} = 2b_1^{22,2}, \quad b_1^{12,2} = 2b_2^{11,2},$$

$$b_2^{12,2} = -b_1^{22,2}, \quad b_1^{21,2} = -b_2^{11,2}.$$

From relation (4.28) for $i = 1$, $j = 1$, $k = 2$, $r = 2$ we obtain

$$b_1^{12,2}b_2^{11,2} + b_2^{12,2}b_2^{21,2} = 0$$

or

$$(b_2^{11,2})^2 - (b_1^{22,2})^2 = 0,$$

that is,

$$b_2^{11,2} = \pm b_1^{22,2}.$$

In what follows we shall consider simultaneously both variants of the sign $\pm$, and in all further formulae the signs $\pm$ and $\mp$ are consistent (the upper sign always corresponds to one case, and the lower sign corresponds to another case if the signs differ).

It follows from identity (4.27) that

$$(c^{si} + c^{is})b_s^{jr,2} = (c^{sj} + c^{js})b_s^{ir,2}.$$

For $r = i = 1$ and $j = 2$ we obtain

$$(c^{11} + c^{22})b_2^{11,2} = (c^{12} + c^{21})b_1^{22,2}.$$

If $b_2^{11,2} = 0$, then all the coefficients $b_k^{ij\,2}$ are equal to zero, and the Poisson structure under consideration is constant. Under the condition $b_2^{11,2} \neq 0$ we obtain the following relation:

$$c^{11} + c^{22} = \pm(c^{12} + c^{21})$$

.

Thus, the second metric $g^{ij\,2}$ either is constant or has the following form in the canonical coordinates w^i:

$$(g^{ij\,2}) = A \begin{pmatrix} 2w^2 & (w^1 \pm w^2) \\ (w^1 \pm w^2) & \pm 2w^1 \end{pmatrix} + \begin{pmatrix} 2a & \pm(a+b) \\ \pm(a+b) & 2b \end{pmatrix},$$

$$\tag{4.32}$$

where $A = b_2^{11,2} \neq 0$, a, b are arbitrary constants.

The metric (4.32) satisfies all relations (4.27)–(4.30).

The constant term in the metric (4.32) can be annihilated by the shift of variables $w^1 = u^1 \mp b/A$, $w^2 = u^2 - a/A$, that is, the corresponding 2-cocycle generated by this constant term on the Lie algebra associated with the Poisson bracket linear with respect to the field variables is cohomologous to zero.

Therefore, our two-dimensional Poisson bracket of hydrodynamic type is reduced to the form

$$
(\{u^i, u^j\}) = \begin{pmatrix} 1 & 0 \\ 0 & -1 \end{pmatrix} \delta_{x^1}(x - y) +
$$

$$
+ A \begin{pmatrix} 2u^2 & (u^1 \pm u^2) \\ (u^1 \pm u^2) & \pm 2u^1 \end{pmatrix} \delta_{x^2}(x - y) + \quad (4.33)
$$

$$
+ A \begin{pmatrix} u^2_{x^2} & 2u^1_{x^2} \mp u^2_{x^2} \\ -u^1_{x^2} \pm 2u^2_{x^2} & \pm u^1_{x^2} \end{pmatrix}.
$$

By Lorentz transformations in $\mathbf{R}^2_1$

$$
\begin{pmatrix} v^1 \\ v^2 \end{pmatrix} = \begin{pmatrix} \tilde{a} & \tilde{b} \\ \tilde{c} & \tilde{d} \end{pmatrix} \begin{pmatrix} u^1 \\ u^2 \end{pmatrix},
$$

where

$$
\tilde{a}^2 - \tilde{b}^2 = 1,
$$

$$
\tilde{c}^2 - \tilde{d}^2 = -1,
$$

$$
\tilde{a}\tilde{c} - \tilde{b}\tilde{d} = 0,
$$

which preserve the Lorentzian metric g^{ij1} of the Poisson bracket (4.33), the Poisson bracket (4.33) is reduced to the canonical form

$$
(\{v^i, v^j\}) = \begin{pmatrix} 1 & 0 \\ 0 & -1 \end{pmatrix} \delta_{x^1}(x - y) +
$$

$$
+ \begin{pmatrix} 2v^2 & (v^1 + v^2) \\ (v^1 + v^2) & 2v^1 \end{pmatrix} \delta_{x^2}(x - y) + \quad (4.34)
$$

$$
+ \begin{pmatrix} v^2_{x^2} & 2v^1_{x^2} - v^2_{x^2} \\ -v^1_{x^2} + 2v^2_{x^2} & v^1_{x^2} \end{pmatrix}.
$$

In fact, after the Lorentz transformation

$$
\tilde{a} = (\text{sign } (\pm A)) \frac{1}{\sqrt{1 - \beta^2}}, \quad \tilde{b} = (\text{sign } (\pm A)) \frac{\beta}{\sqrt{1 - \beta^2}},
$$

$$\tilde{c} = \pm(\text{sign } (\pm A)) \frac{\beta}{\sqrt{1 - \beta^2}}, \qquad \tilde{d} = \pm(\text{sign } (\pm A)) \frac{1}{\sqrt{1 - \beta^2}},$$

where $-1 < \beta < 1$, the metric $g^{ij\,2}$ of the Poisson bracket (4.33) becomes

$$(g^{ij\,2}) = \frac{\sqrt{1 - \beta^2}}{1 \mp \beta}|A| \begin{pmatrix} 2v^2 & (v^1 + v^2) \\ (v^1 + v^2) & 2v^1 \end{pmatrix}.$$

Thus, since for any $A \neq 0$ there exists a number β, $-1 < \beta < 1$, such that

$$\sqrt{\frac{1 - \beta}{1 + \beta}}|A| = 1$$

or, respectively,

$$\sqrt{\frac{1 + \beta}{1 - \beta}}|A| = 1,$$

Theorem 4.4 is proved.

For the canonical Poisson structure generated by the metrics (4.26) the obstruction tensor $T^{ijk,12}$ is not equal to zero, in particular, $T^{211,12} = -b_2^{11,2} = -1$, and consequently this Poisson bracket cannot be reduced to a constant form by local changes of variables.

Proposition 4.1 ([105]) *The Poisson structure generated by the Lie algebra of vector fields on T^2:*

$$\{w^i(x), w^j(y)\} = w^i(x)\delta_j(x - y) + w^j(x)\delta_i(x - y) + w_j^i(x)\delta(x - y),$$

reduces to the canonical form by the local quadratic change of coordinates

$$\begin{cases} w^1 = \frac{1}{2}((v^1)^2 - (v^2)^2), \\ w^2 = \frac{1}{2}(v^1 + v^2). \end{cases}$$

This example is connected with the two-dimensional Euler hydrodynamics of an ideal incompressible fluid (with further reduction to divergent-free vector fields).

Corollary 4.2 ([105]) *Any two-dimensional two-component non-degenerate Poisson structure of hydrodynamic type either can be reduced to a constant form or is generated by the Lie algebra of vector fields on T^2.*

Theorem 4.5 ([105]) *For $n > 2$ an arbitrary multidimensional two-component non-degenerate Poisson structure of hydrodynamic type reduces either to a constant form by a local change of coordinates or to a two-dimensional Poisson structure by an unimodular change of the independent space variables x^i.*

4.2 Homogeneous Hamiltonian systems of hydrodynamic type and metrics of constant Riemannian curvature

4.2.1 One-dimensional homogeneous Hamiltonian systems of hydrodynamic type

Let us consider in more detail the *one-dimensional homogeneous systems of hydrodynamic type*, that is, evolution quasilinear systems of first-order partial differential equations

$$u_t^i = v_j^i(u)u_x^j, \qquad (4.35)$$

where $v_j^i(u)$ is an arbitrary $N \times N$ matrix function of N variables, $u = (u^1, \ldots, u^N)$, $u^i = u^i(x,t)$, $i = 1, \ldots, N$.

Dubrovin and Novikov introduced and described the class of one-dimensional homogeneous Hamiltonian systems of hydrodynamic type

$$u_t^i = \{u^i(x), H\}, \qquad (4.36)$$

where H is an arbitrary functional of hydrodynamic type, that is,

$$H = \int h(u(x))\,dx, \qquad (4.37)$$

and the Poisson bracket has the form

$$\{u^i(x), u^j(y)\} = g^{ij}(u(x))\delta_x(x - y) + b_k^{ij}(u(x))u_x^k\delta(x - y) \quad (4.38)$$

(the Dubrovin–Novikov bracket). The form of the Poisson bracket (4.38) is invariant with respect to local changes of coordinates on the manifold M, and the coefficients of the bracket are transformed like differential-geometric objects on M. The Dubrovin–Novikov theorem gives a complete description of non-degenerate

Poisson brackets of the form (4.38) (see Section 4.1). The Poisson structures (4.38) are invariant with respect to the group $\mathrm{Diff}^+(S^1)$ of orientation-preserving diffeomorphisms of the circle S^1, that is, they does not depend on the parametrizations of the loops. Non-degenerate Poisson structures (4.38) are defined on the loop spaces of arbitrary flat manifolds.

All these Hamiltonian systems of hydrodynamic type have a natural Lagrangian origin. Let us consider the following action of type (2.104):

$$S = \int \left(\frac{1}{2} g_{ij} u_x^i u_t^j - h(u_x) \right) dx \, dt,$$

where (g_{ij}) is a constant non-degenerate symmetric metric, and $h(v)$ is an arbitrary function. The corresponding Lagrangian system $\delta S/\delta u^i(x) = 0$ has a symplectic representation of the form

$$g_{ij} \frac{d}{dx}(u_t^j) = -\frac{\delta H}{\delta u^i(x)}, \qquad H = \int h(u_x) \, dx.$$

After the transformation $v^i(x) = u_x^i(x)$ we obtain a Hamiltonian system of hydrodynamic type in flat coordinates

$$v_t^i = g^{ij} \frac{d}{dx} \frac{\delta H}{\delta v^j(x)}, \qquad H = \int h(v(x)) \, dx. \qquad (4.39)$$

According to Dubrovin–Novikov theorem any one-dimensional Hamiltonian system of hydrodynamic type with a non-degenerate Poisson bracket of hydrodynamic type reduces to the form (4.39) by a local change of coordinates.

Thus, the action S generates the general class of homogeneous Hamiltonian systems of hydrodynamic type (written in flat coordinates $(v^1, \ldots, v^N)$) introduced and studied by Dubrovin and Novikov [34].

The homogeneous Hamiltonian systems of hydrodynamic type under consideration can be rewritten in the form

$$u_t^i = [\nabla^i \nabla_j h(u)] u_x^j, \qquad (4.40)$$

where ∇ is the covariant derivative generated by a metric of zero curvature.

It was proved by Tsarev [178], [181] that if a homogeneous Hamiltonian system of hydrodynamic type (4.40) possesses the complete set of Riemann invariants, that is, the matrix

$$v^i_j(u) = \nabla^i \nabla_j h(u)$$

is diagonalizable, then this system is integrable by the generalized hodograph method.

We recall that the function $R(u^1, ..., u^N)$ is called a *Riemann invariant* of the system of hydrodynamic type (4.35) if there is a function $\alpha(u^1, ..., u^N)$ such that by virtue of the system (4.35) we have

$$R_t = \alpha(u)R_x.$$

In particular, a system of hydrodynamic type is diagonalizable if and only if this system possesses the complete set of Riemann invariants (in other words, the set of N different and functionally independent Riemann invariants which can be taken as new coordinates $\{R^1, ..., R^N\}$, $R^i = R^i(u^1, ..., u^N)$, $i = 1, ..., N$):

$$R^i_t = \alpha^i(R)R^i_x, \qquad i = 1, ..., N. \qquad (4.41)$$

The diagonal system (4.41) is called *semi-Hamiltonian* if its coefficients $\alpha^i(R)$, $i = 1, ..., N$, satisfy the following relations ([178], [181]):

$$\frac{\partial}{\partial R^i}\left(\frac{1}{\alpha^j - \alpha^k}\frac{\partial\alpha^k}{\partial R^j}\right) = \frac{\partial}{\partial R^j}\left(\frac{1}{\alpha^i - \alpha^k}\frac{\partial\alpha^k}{\partial R^i}\right), \qquad (4.42)$$
$$i \neq j, \quad i \neq k, \quad j \neq k.$$

In particular, Tsarev proved that any diagonal Hamiltonian system of hydrodynamic type is always semi-Hamiltonian [178], [181].

Theorem 4.6 (Tsarev [178], [181]) *Let the system (4.41) be a diagonal Hamiltonian (or semi-Hamiltonian) system of hydrodynamic type. If $w^1(R), ..., w^N(R)$ is an arbitrary solution of the linear system*

$$\frac{1}{w^i - w^k}\frac{\partial w^k}{\partial R^i} = \frac{1}{\alpha^i - \alpha^k}\frac{\partial\alpha^k}{\partial R^i}, \qquad i \neq k, \qquad (4.43)$$

then the functions $R^1(x,t), ..., R^N(x,t)$ *determined by the system of equations*

$$w^i(R) = \alpha^i(R)t + x, \quad i = 1, ..., N, \qquad (4.44)$$

satisfy the system of hydrodynamic type (4.41).

Moreover, every smooth solution of this system is always locally obtainable in this way.

The relations (4.42) are equivalent precisely to the compatibility conditions for the linear system (4.43).

4.2.2   Non-local Poisson brackets of hydrodynamic type related to metrics of constant Riemannian curvature

The author of the present paper and Ferapontov [151] proposed a non-local generalization of the Hamiltonian theory of one-dimensional systems of hydrodynamic type (4.35). This generalization is connected with non-local Poisson brackets of the form (the Mokhov–Ferapontov brackets)

$$\{u^i(x), u^j(y)\} = g^{ij}(u(x))\delta_x(x-y) + \qquad (4.45)$$

$$+ b_k^{ij}(u(x))u_x^k\delta(x-y) + Ku_x^i\left(\frac{d}{dx}\right)^{-1}u_x^j\delta(x-y),$$

where K is an arbitrary constant.

For any Hamiltonian H of hydrodynamic type (4.37) and for any non-local Poisson bracket of the form (4.45) we shall always obtain a system of hydrodynamic type (4.35). Moreover, (4.45) is the most general form of Poisson brackets having the property of generating systems of hydrodynamic type (4.35) for any Hamiltonian H of hydrodynamic type (4.37) (if the number of components $N > 1$; of course, in the one-component case it is possible explicitly to point out different integro-differential operators of arbitrary odd orders that possess this property).

Lemma 4.2 ([121]) *The expression (4.45) defines a Poisson bracket if and only if the following relations are fulfilled:*

$$g^{ij} = g^{ji}, \tag{4.46}$$

$$\frac{\partial g^{ij}}{\partial u^k} = b_k^{ij} + b_k^{ji}, \tag{4.47}$$

$$g^{is} b_s^{jr} = g^{js} b_s^{ir}, \tag{4.48}$$

$$g^{is} \left(\frac{\partial b_s^{jr}}{\partial u^k} - \frac{\partial b_k^{jr}}{\partial u^s} \right) + b_s^{ij} b_k^{sr} - b_s^{ir} b_k^{sj} = K(g^{ir} \delta_k^j - g^{ij} \delta_k^r), \tag{4.49}$$

$$\sum_{(k,p)} \sum_{(i,j,r)} \left[b_p^{si} \left(\frac{\partial b_k^{jr}}{\partial u^s} - \frac{\partial b_s^{jr}}{\partial u^k} \right) + K(b_k^{ij} - b_k^{ji}) \delta_p^r \right] = 0. \tag{4.50}$$

In Lemma 4.2 we do not assume that the matrix g^{ij} is non-degenerate. The general relations (4.46)–(4.50) are obtained by direct calculations from the Jacobi identity and skew-symmetry of the Poisson structure. Relations (4.46) and (4.47) are equivalent to skew-symmetry of the bracket (4.45), and relations (4.48)–(4.50) are equivalent to the fulfilment of the Jacobi identity for a skew-symmetric bracket (4.45).

Theorem 4.7 ([151]) *If*

$$\det[g^{ij}(u)] \neq 0,$$

then the expression (4.45) defines a Poisson bracket if and only if

1. $g^{ij}(u)$ *is a metric of constant Riemannian curvature K,*

2. $b_k^{ij}(u) = -g^{is} \Gamma_{sk}^j(u)$, *where $\Gamma_{sk}^j(u)$ are the coefficients of the Riemannian connection generated by the metric $g^{ij}(u)$ (the Levi–Civita connection).*

Proof. It follows from (4.46) and the transformation law for the coefficients of the Poisson structure that g^{ij} is a symmetric Riemannian or pseudo-Riemannian metric on the manifold. Introducing the coefficients of differential-geometric connection Γ_{jk}^i by the formula

$$b_k^{ij}(u) = -g^{is}(u) \Gamma_{sk}^j(u),$$

we deduce that relation (4.47) is equivalent to the condition that
the connection is compatible with the metric g^{ij}:

$$\nabla_k g^{ij} \equiv \frac{\partial g^{ij}}{\partial u^k} + g^{sj}\Gamma^i_{sk} + g^{is}\Gamma^j_{sk} = 0.$$

Relation (4.48) is equivalent to the condition that the connection
is symmetric:

$$T^i_{jk}(u) \equiv \Gamma^i_{jk}(u) - \Gamma^i_{kj}(u) = 0.$$

Thus, Γ^i_{jk} is the Levi–Civita connection, that is, the Riemannian
connection generated by the metric. If these conditions are fulfilled,
then relation (4.49) can be rewritten as follows:

$$R^i_{jks}(u) = K(g_{jk}(u)\delta^i_s - g_{js}(u)\delta^i_k), \qquad (4.51)$$

where $R^i_{jks}(u)$ is the Riemannian curvature tensor. Relation (4.51)
for the Riemannian curvature tensor means precisely that the metric $g^{ij}(u)$ is a metric of constant Riemannian curvature K. Relation (4.50) is a consequence of relations (4.46)–(4.49) in the case
of a non-degenerate metric $g^{ij}(u)$ and therefore it is automatically
fulfilled for the metric of constant Riemannian curvature (in the
case of a degenerate metric, this relation is non-trivial).

Note that in the one-component case (for $N = 1$) relation (4.51)
is fulfilled for any value of the constant K, that is, if condition (4.47)
is fulfilled, then (4.45) defines a one-component Poisson structure
for any K. We do not explicitly specify this trivial case in Theorem 4.7, assuming conditionally that it is possible to attribute
any value of curvature K to any one-component metric. It is interesting that the examples of the Liouville equation and the sine-
Gordon equation (see Section 2.2, formulae (2.98), (2.99), (2.101))
differ from each other namely by the sign of the "conditional curvature" K. Along with the Liouville and sine-Gordon equations
and the systems of hydrodynamic type, a non-local operator (or
the non-local part (the *non-local "tail"*) of an operator) of the
form $u^i_x(d/dx)^{-1} \circ u^j_x$ naturally arises in the Hamiltonian structure
of the Krichever–Novikov equation (see formulae (2.123)–(2.125)
in Section 2.3), in the third Hamiltonian structure of the Korteweg–
de Vries equation (see formula (2.119) in Section 2.3), and in the
Hamiltonian theory of the Heisenberg ferromagnets (see Section
4.4).

The Hamiltonian systems of hydrodynamic type

$$u_t^i = \{u^i(x), H\}, \quad H = \int h(u(x))dx,$$

that correspond to the non-degenerate Poisson structures (4.45) can be rewritten in the form

$$u_t^i = [\nabla^i \nabla_j h(u) + K \delta_j^i h(u)] u_x^j,$$

where ∇ is the covariant derivative generated by a metric of constant Riemannian curvature K.

Any diagonal system of hydrodynamic type which is Hamiltonian with respect to the non-degenerate Poisson brackets (4.45) is also always semi-Hamiltonian.

On local translation-invariant functionals of the form

$$F = \int f(u, u_x, \ldots, u_{(k)})\,dx$$

the Poisson structure (4.45) always generates a local expression, since $u_x^i \frac{\delta F}{\delta u^i(x)}$ is a total derivative with respect to x:

$$\int_{S^1} \frac{df}{dx}\,dx = \int_{S^1} \left[\frac{\partial f}{\partial u^i} u_x^i + \frac{\partial f}{\partial u_x^i} u_{xx}^i + \right.$$
$$\left. + \cdots + \frac{\partial f}{\partial u_{(k)}^i} u_{(k+1)}^i \right] dx = \int_{S^1} u_x^i \frac{\delta F}{\delta u^i(x)} = 0.$$

In particular, on functionals of hydrodynamic type

$$G_i = \int g_i(u)\,dx, \quad i = 1, 2,$$

on the loop space ΩM of a manifold (M, g_{ij}) of constant Riemannian curvature K the non-degenerate Poisson bracket (4.45) can be rewritten in the following local form:

$$\{G_1, G_2\} = \int_{S^1} (\langle \nabla g_1, \nabla_\nu \nabla g_2 \rangle + K g_2 \nabla_\nu g_1)\,dx,$$

where ∇ is the covariant derivative generated by the metric of constant Riemannian curvature K, $\nu = \{u_x^i\}$ is the velocity vector field of the loop $\gamma(x)$, ∇_ν is the covariant derivative along the loop $\gamma(x)$, and

$$\langle \nabla h, \nabla f \rangle = g^{ij} \nabla_i h \nabla_j f$$

is the scalar product of covector fields on M.

The non-local Poisson brackets of hydrodynamic type under consideration are defined on the loop spaces ΩM of manifolds M of constant Riemannian curvature K and they are invariant with respect to the action of the group $\mathrm{Diff}^+(S^1)$ of orientation-preserving diffeomorphisms of the circle S^1.

The canonical form of the non-local Poisson brackets with respect to local changes of coordinates on M is defined by the canonical metrics of constant Riemannian curvature K:

$$(g^{ij}(u)) = [\lambda(u)]^2 \begin{pmatrix} e_1 & 0 & \ldots & 0 \\ 0 & e_2 & \ldots & 0 \\ \vdots & \vdots & \ddots & \vdots \\ 0 & 0 & \ldots & e_N \end{pmatrix}, \qquad (4.52)$$

$$e_i = \pm 1, \qquad \lambda(u) = 1 + \frac{1}{4} K \sum_i e_i (u^i)^2.$$

It is obvious that they cannot be reduced to a constant form by local changes of coordinates on M.

Note that the homogeneous symplectic structures of the first order considered in Section 2 do not possess any non-local generalizations similar to (4.45), that is, of the form

$$\omega(\xi, \eta) = \int_{S^1} \xi^i \left(g_{ij}(u)\eta_x^j + b_{ijk}(u)u_x^k \eta^j + \right.$$

$$\left. + w_{is}(u)u_x^s \left(\frac{d}{dx} \right)^{-1} w_{jr}(u)u_x^r \eta^j \right) dx.$$

If $\det[g^{ij}(u)] = 0$, then the geometrical description of non-local Poisson brackets (4.45) is very complicated (the corresponding relations (4.46)–(4.50) were obtained by the present author in [121]). A complete geometrical description has not been obtained even in the local case (for $K = 0$); see [75].

The Poisson structures connected with the metrics of constant Riemannian curvature have numerous applications in mathematical physics. We consider here the simplest example of the Chaplygin gas equations.

Example 4.2 Let us consider the system of Chaplygin gas equations

$$\begin{cases} U_t = UU_x + V^{-3}V_x, \\ V_t = VU_x + UV_x. \end{cases} \tag{4.53}$$

We rewrite this system in the Riemann invariants u, v:

$$\begin{cases} u_t = vu_x, \\ v_t = uv_x, \end{cases} \tag{4.54}$$

where

$$u = U - \frac{1}{V}, \qquad v = U + \frac{1}{V}.$$

This system possesses the Poisson structures of hydrodynamic type generated by diagonal metrics g^{ij} of the following form:

$$g^{11} = -[(c_1 + K) + c_2 u + c_3 u^2](u - v)^2,$$

$$g^{22} = [c_1 + c_2 v + c_3 v^2](u - v)^2.$$

For any constants c_1, c_2, c_3 we obtain a metric of constant Riemannian curvature K, generating the corresponding non-local Poisson structure of the Chaplygin gas equations. For $K = 0$ we obtain flat metrics corresponding to three different local Poisson structures of hydrodynamic type for the Chaplygin gas equations. All these local and non-local Poisson structures are compatible, that is, they generate a linear pencil of Poisson structures.

In [38] Ferapontov classified all two-component systems of hydrodynamic type possessing one non-local and three local Poisson structures of hydrodynamic type.

Example 4.3 ([39], [53]) Let us consider a diagonal system of hydrodynamic type

$$R^i_t = \left(\sum_{k=1}^{n} R^k + 2R^i \right) R^i_x,$$

which describes the quasiclassical limit of coupled Korteweg–de Vries equations [2]–[4], [39], [53], [57].

This system possesses $n + 1$ local and one non-local Poisson structure of hydrodynamic type, which are compatible. The flat diagonal metrics

$$(ds^a)^2 = \sum_{i=1}^{n} g_{ii}^a (dR^i)^2 =$$

$$= \sum_{i=1}^{n} \left(\frac{\prod_{k \neq i}(R^k - R^i)}{(R^i)^a} \right) (dR^i)^2, \qquad a = 0, 1, \ldots, n,$$

generate compatible local Poisson structures of the system under consideration:

$$(A^a)^{ij} = (g^a)^{ii} \delta^{ij} \frac{d}{dx} - (g^a)^{ii} \Gamma_{ik}^j R_x^k, \qquad a = 0, 1, \ldots, n,$$

where Γ_{ik}^j is the Levi–Civita connection corresponding to the metric g_{ii}^a.

The recursion operator $R = A^1(A^0)^{-1}$ has the form:

$$R = \begin{pmatrix} R^1 & 0 & \cdots & 0 \\ 0 & R^2 & \cdots & 0 \\ \vdots & \vdots & \ddots & \vdots \\ 0 & 0 & \cdots & R^n \end{pmatrix} +$$

$$+ \frac{1}{2} \begin{pmatrix} R_x^1 & R_x^1 & \cdots & R_x^1 \\ R_x^2 & R_x^2 & \cdots & R_x^2 \\ \vdots & \vdots & \ddots & \vdots \\ R_x^n & R_x^n & \cdots & R_x^n \end{pmatrix} \left(\frac{d}{dx} \right)^{-1}. \qquad (4.55)$$

For all $a = 0, 1, \ldots, n - 1$ the following recursion relation is fulfilled: $A^{a+1} = RA^a$. The operator $A^{n+1} = RA^n$ is non-local and has the form:

$$(A^{n+1})^{ij} = (g^{n+1})^{ii} \delta^{ij} \frac{d}{dx} -$$

$$- (g^{n+1})^{ii} \Gamma_{ik}^j R_x^k + \frac{1}{4} R_x^i \left(\frac{d}{dx} \right)^{-1} \circ R_x^j, \qquad (4.56)$$

where

$$g_{ii}^{n+1} = \frac{\prod_{k \neq i}(R^k - R^i)}{(R^i)^{n+1}}$$

is a metric of constant Riemannian curvature $1/4$ generating the non-local Poisson structure A^{n+1}.

Essentially, as was shown in [53], for $n \geq 3$ all systems of hydrodynamic type possessing $n + 1$ local and one non-local Poisson structure of hydrodynamic type reduce to the hierarchy of the system under consideration, that is, the quasiclassical limit of coupled KdV equations, which is written here above in the Riemann invariants.

The Poisson structures generated by the metrics of constant Riemannian curvature also have very natural applications in the theory of non-homogeneous systems of hydrodynamic type and in the theory of the Heisenberg ferromagnets (see Section 4.4).

A number of important concrete examples of non-local Poisson brackets of hydrodynamic type (4.45) arising in mathematical physics, in particular, as Hamiltonian structures of the Whitham equations obtained by averaging the Korteweg–de Vries equation, the non-linear Schrödinger equation, and the sine-Gordon equation, were later explicitly described and studied by Pavlov and Alekseev in [1], [169], [170].

Example 4.4 ([1]) Let us consider a family of diagonal metrics of the form

$$(g_n^{ii})^{-1} = \frac{(g_0^{ii})^{-1}}{(u^i)^n}, \quad i = 1, \dots, 2g + 1, \quad n = 1, 2, \dots,$$

$$(g_0^{ii})^{-1} = 2 \operatorname*{res}_{\lambda = u^i} \left[\frac{dp}{d\lambda} \right]^2 d\lambda, \tag{4.57}$$

where dp is the differential of quasimomentum arising in the theory of the KdV equation, that is, the Abelian differential of the second kind on the Riemann surface

$$\mu^2 = \prod_{i=1}^{2g+1} (\lambda - u^i) \tag{4.58}$$

with the only pole at the infinite point $\lambda = \infty$ of the form

$$dp = d(-\sqrt{\lambda} + O(1))$$

and with the normalization conditions

$$\int_{u^{2i}}^{u^{2i+1}} dp = 0, \qquad i = 1, \ldots, g. \tag{4.59}$$

Here all u^i are real and distinct: $u^1 < u^2 < \cdots < u^{2g+1}$. All the constructed metrics are non-degenerate. The metrics g_0^{ii} and g_1^{ii} are flat metrics generating two local compatible Poisson structures of hydrodynamic type A_0^{ij} and A_1^{ij} for the Whitham equations describing slow modulations of g-gap solutions of the KdV equation:

$$u_t^i = w^i(u^1, \ldots, u^{2g+1})u_x^i, \qquad i = 1, 2, \ldots, 2g+1, \tag{4.60}$$

where $w^i = dq/dp$ ($\lambda = u^i$), dq is the Abelian differential of the second kind on the Riemann surface (4.58) with singularity of the form

$$dq = d(-(\sqrt{\lambda})^3 + O(1))$$

at the infinite point and with normalization similar to (4.59). Defining the recursion operator R from the relation $A_1^{ij} = R A_0^{ij}$, we consider the Poisson structures $A_n^{ij} = R^n A_0^{ij}$. For $n = 2$ we obtain non-local Poisson structures of hydrodynamic type A_2^{ij} generated by the metric g_n^{ii} of constant Riemannian curvature $-1/4$. For $n > 2$ the Poisson structures A_n^{ij} have more complicated non-local form:

$$A_n^{ij} = g_n^{ii}\delta^{ij}\frac{d}{dx} - g_n^{ii}\Gamma_{ik}^j u_x^k -$$

$$-\frac{1}{4}\sum_{l=1}^{n-1}(w^l)^i u_x^i \left(\frac{d}{dx}\right)^{-1} \circ (w^{n-l})^j u_x^j, \tag{4.61}$$

where $(w^l)^i = ds^l/dp$ ($\lambda = u^i$), ds^l is the Abelian differential of the second kind on the Riemann surface (4.58) with singularity of the form

$$ds^l = d(-(\sqrt{\lambda})^{2l-1} + O(1))$$

and with normalization similar to (4.59).

4.2.3   Further non-local generalizations of Poisson brackets of hydrodynamic type

The differential geometry of non-local Poisson structures of the type (4.61), which are a generalization of the non-local Poisson structures of hydrodynamic type, was studied by Ferapontov in [39] (the Ferapontov brackets).

Theorem 4.8 ([39]) *The expression*

$$\{u^i(x), u^j(y)\} = g^{ij}(u(x))\delta_x(x - y) + \qquad (4.62)$$
$$+b^{ij}_k(u(x))u^k_x\delta(x - y) +$$
$$+w^i_k(u)u^k_x\left(\frac{d}{dx}\right)^{-1} w^j_s(u)u^s_x\delta(x - y), \quad \det[g^{ij}(u)] \neq 0,$$

defines a Poisson bracket if and only if

1. $b^{ij}_k(u) = -g^{is}(u)\Gamma^j_{sk}(u)$, *where* $\Gamma^j_{sk}(u)$ *are the coefficients of the affine connection generated by a symmetric (pseudo-Riemannian) metric* $g^{ij}(u)$ *(the Levi–Civita connection),*

2. *the pseudo-Riemannian metric* $g_{ij}(u)$ *(with lower indices) and the affinor* $w^i_j(u)$ *satisfy the Gauss–Peterson–Codazzi relations:*

$$g_{ik}w^k_j = g_{jk}w^k_i, \qquad \nabla_k w^i_j = \nabla_j w^i_k,$$
$$R^{ij}_{kl} = w^i_k w^j_l - w^j_k w^i_l, \qquad (4.63)$$

where $R^{ij}_{kl} \equiv g^{is}R^j_{skl}$ *is the Riemannian curvature tensor.*

Thus, the classic Gauss–Peterson–Codazzi relations for hypersurfaces M of a pseudo-Euclidean space E^{N+1} are equivalent to the skew-symmetry and the fulfilment of the Jacobi identity for the non-local Poisson bracket (4.62). Here g_{ij} is the first fundamental form of the hypersurface M and the affinor w^i_j is the Weingarten operator (shape operator). The Dubrovin–Novikov brackets correspond to hyperplanes in E^{N+1}. In this case the Weingarten operator vanishes. In the case of a hypersphere of radius 1, $w^i_j = \delta^i_j$, and we obtain the non-local Poisson bracket of hydrodynamic type (4.45).

It is important to note that in the case of the Poisson brackets
(4.62), in contrast to (4.45), it is necessary specially to describe the
class of Hamiltonians H of hydrodynamic type (4.37), which generate systems of hydrodynamic type (4.35), since this condition is far
from being fulfilled for all of them: the functional $H = \int h(u)\,dx$
must be a first integral of the system

$$u_t^i = w_j^i(u)u_x^j,$$

where $w_j^i(u)$ is the affinor from (4.62) (see [39]).

Further non-local generalization of the Dubrovin–Novikov brackets (4.38) leads to N-dimensional surfaces with flat normal connections, which are embedded in a pseudo-Euclidean space E^{N+L}
[39]. We recall that an N-dimensional surface M in a pseudo-Euclidean space E^{N+L} is called a *surface with flat normal connection* if
it possesses a framing by the field of unit normals $\vec{n}^\alpha$, $\alpha = 1, \dots, L$,
such that $d\vec{n}^\alpha \in TM$.

Theorem 4.9 ([39]) *The expression*

$$\{u^i(x), u^j(y)\} = g^{ij}(u(x))\delta_x(x-y) + \tag{4.64}$$
$$+b_k^{ij}(u(x))u_x^k\delta(x-y) +$$
$$+ \sum_{\alpha=1}^{L} (w^\alpha)_k^i(u)u_x^k\left(\frac{d}{dx}\right)^{-1}(w^\alpha)_s^j(u)u_x^s\delta(x-y), \quad \det[g^{ij}(u)] \neq 0,$$

defines a Poisson bracket if and only if

1. $b_k^{ij}(u) = -g^{is}(u)\Gamma_{sk}^j(u)$, *where* $\Gamma_{sk}^j(u)$ *are the coefficients*
   of the affine connection generated by a symmetric (pseudo-
   Riemannian) metric $g^{ij}(u)$ *(the Levi–Civita connection),*

2. *the pseudo-Riemannian metric* $g_{ij}(u)$ *(with lower indices)*
   and the set of affinors $(w^\alpha)_j^i(u)$ *satisfy the relations:*

$$g_{ik}(w^\alpha)_j^k = g_{jk}(w^\alpha)_i^k, \quad \nabla_k(w^\alpha)_j^i = \nabla_j(w^\alpha)_k^i, \tag{4.65}$$
$$R_{kl}^{ij} = \sum_{\alpha=1}^{L}\{(w^\alpha)_k^i(w^\alpha)_l^j - (w^\alpha)_k^j(w^\alpha)_l^i\}, \quad \alpha = 1, \dots, L.$$

Moreover, the family of affinors w^α *is commutative:*

$$[w^\alpha, w^\beta] = 0.$$

Relations (4.65) are nothing but the Gauss–Peterson–Codazzi equations for N-dimensional surfaces M with flat normal connections in a pseudo-Euclidean space E^{N+L}. Here g_{ij} is the first fundamental form of the surface M, and w^α are the Weingarten operators corresponding to the field of normals $\vec{n}^\alpha$.

All local and non-local Poisson brackets of hydrodynamic type, considered in the present section, are always defined on the loop spaces ΩM of the corresponding manifolds M (in particular, for (4.45) they are manifolds of constant Riemannian curvature K, for (4.62) they are arbitrary hypersurfaces in a pseudo-Euclidean space E^{N+1}, for (4.64) they are arbitrary surfaces with flat normal connections embedded in a pseudo-Euclidean space E^{N+L}), and all of them are invariant with respect to the action of the group $\mathrm{Diff}^+(S^1)$ of orientation-preserving diffeomorphisms of the circle S^1.

In the next section we shall also consider a non-homogeneous generalization of local and non-local Poisson structures of hydrodynamic type.

4.3 Non-homogeneous Hamiltonian systems of hydrodynamic type

4.3.1 Non-homogeneous local multidimensional Poisson brackets of hydrodynamic type

Let us consider *non-homogeneous multidimensional systems of hydrodynamic type*

$$u_t^i = v_j^{i\alpha}(u)\frac{\partial u^j}{\partial x^\alpha} + f^i(u). \qquad (4.66)$$

Non-homogeneous local multidimensional Poisson structures of hydrodynamic type are defined by the formula

$$\{u^i(x), u^j(y)\} = g^{ij\alpha}(u(x))\delta_{x^\alpha}(x - y) +$$
$$+ b_k^{ij\alpha}(u(x))u_{x^\alpha}^k\delta(x - y) + \omega^{ij}(u(x))\delta(x - y). \qquad (4.67)$$

Lemma 4.3 *The expression (4.67) defines a Poisson structure if and only if the right-hand side of (4.67) is the sum of a homogeneous multidimensional Poisson structure of hydrodynamic type*

118   O.I. MOKHOV

(see formula (4.4) in Section 4.1) and a classic finite-dimensional Poisson structure $\omega^{ij}(u)$, and these Poisson structures are compatible, that is, the following compatibility conditions are fulfilled:

$$\Phi^{ijk,\alpha} = \Phi^{kij,\alpha}, \qquad (4.68)$$

where

$$\Phi^{ijk,\alpha} = g^{is\alpha}\frac{\partial \omega^{jk}}{\partial u^s} - b_s^{ij\alpha}\omega^{sk} - b_s^{ik\alpha}\omega^{js};$$

$$\frac{\partial \Phi^{ijk,\alpha}}{\partial u^r} = \sum_{(i,j,k)}\left[b_r^{si\alpha}\frac{\partial \omega^{jk}}{\partial u^s} + \right.$$
$$\left. + \left(\frac{\partial b_r^{ij\alpha}}{\partial u^s} - \frac{\partial b_s^{ij\alpha}}{\partial u^r}\right)\omega^{sk}\right]. \qquad (4.69)$$

In Lemma 4.3 non-degeneracy of the metrics is not assumed. For the homogeneous multidimensional Poisson bracket of hydrodynamic type which is a part of the bracket (4.67) relations (4.5)–(4.11) must be fulfilled, and for the finite-dimensional Poisson structure $\omega^{ij}(u)$ the following classic relations are satisfied:

$$\omega^{ij} = -\omega^{ji}, \qquad (4.70)$$

$$\omega^{is}\frac{\partial \omega^{jk}}{\partial u^s} + \omega^{js}\frac{\partial \omega^{ki}}{\partial u^s} + \omega^{ks}\frac{\partial \omega^{ij}}{\partial u^s} = 0. \qquad (4.71)$$

Lemma 4.4 *If the metrics $g^{ij\alpha}$ are non-degenerate, then the compatibility conditions (4.68) and (4.69) have the following form:*

$$\nabla^{i,\alpha}\omega^{jk} + \nabla^{j,\alpha}\omega^{ik} = 0, \qquad (4.72)$$

$$\nabla_s^\alpha \nabla_k^\alpha \omega^{ij} = 0, \qquad (4.73)$$

where ∇_s^α is the covariant derivative generated by the metric $g^{ij\alpha}$.

Thus, the compatibility conditions have a clear differential-geometric sense: in particular, condition (4.72) means precisely that a skew-symmetric tensor $\omega^{ij}(u)$ is a Killing bivector for each of the flat Riemannian or pseudo-Riemannian metrics $g^{ij\alpha}$.

We recall that a skew-symmetric tensor $\xi_{i_1 i_2 \ldots i_p}$ on a Riemannian or pseudo-Riemannian manifold (M, g_{ij}) is called a *Killing tensor* if it satisfies the relation

$$\nabla_k \xi_{i i_2 \ldots i_p} + \nabla_i \xi_{k i_2 \ldots i_p} = 0,$$

where ∇_i is the Riemannian connection generated by the metric. Killing tensors are specified by the following property: for a skew-symmetric tensor $\xi_{i_1 i_2 \ldots i_p}$ the tensor $\xi_{i i_2 \ldots i_p} u_x^i$ is covariantly constant along any geodesic $u^i(x)$ on the manifold M if and only if the tensor $\xi_{i_1 i_2 \ldots i_p}$ is a Killing tensor.

In fact, if $u^i(x)$ is an arbitrary geodesic on M, then

$$\nabla_k (\xi_{i i_2 \ldots i_p} u_x^i) = \frac{1}{2}(\nabla_k \xi_{i i_2 \ldots i_p} + \nabla_i \xi_{k i_2 \ldots i_p}) u_x^i u_x^k = 0.$$

In particular, for $p = 1$ we obtain the classic Killing vectors.

Killing tensors also possess the following fundamental property: for them the covariant derivative tensor $\nabla_k \xi_{i_1 i_2 \ldots i_p}$ is skew-symmetric with respect to any pair of indices. Arbitrary skew-symmetric tensors whose covariant derivatives are also skew-symmetric tensors were considered for the first time by Bochner (see [10], [191], [193]). In the present survey we shall be interested in contravariant Killing bivectors b^{ij} with indices raised with the help of the corresponding metrics.

4.3.2   Killing–Poisson bivectors on flat manifolds and Lie–Poisson bivectors

If a Killing bivector on a Riemannian or pseudo-Riemannian manifold (M, g_{ij}) is simultaneously a Poisson bivector, then it is called a *Killing–Poisson bivector*.

Theorem 4.10 *An ultralocal Poisson structure $\omega^{ij}(u)$ is compatible with the non-degenerate multidimensional homogeneous Poisson structure of hydrodynamic type (4.4) if and only if it is a Killing–Poisson bivector for each of the metrics $g^{ij\alpha}$.*

By Lemmas 4.3 and 4.4 it is sufficient to prove that relation (4.73) is fulfilled for Killing bivectors on the corresponding flat manifolds. In fact,

$$\nabla^{k,\alpha}\nabla^{s,\alpha}\omega^{ij} = -\nabla^{k,\alpha}\nabla^{i,\alpha}\omega^{sj} =$$
$$= -\nabla^{i,\alpha}\nabla^{k,\alpha}\omega^{sj} = \nabla^{i,\alpha}\nabla^{s,\alpha}\omega^{kj} =$$
$$= \nabla^{s,\alpha}\nabla^{i,\alpha}\omega^{kj} = -\nabla^{s,\alpha}\nabla^{k,\alpha}\omega^{ij} = -\nabla^{k,\alpha}\nabla^{s,\alpha}\omega^{ij},$$

whence it follows immediately that $\nabla_k^\alpha \nabla_s^\alpha \omega^{ij} = 0$. We use the fact that in our case all covariant derivatives ∇_k^α are generated by the flat metrics $g^{ij\alpha}$. Since all the metrics $g^{ij\alpha}$ are flat, we can consider coordinates w^i in which one of the metrics, for example g^{ij1}, is constant. In these flat coordinates all Christoffel symbols Γ_{jk}^{i1} vanish. From condition (4.73) it follows immediately that the bivector ω^{ij} is linear in these coordinates:

$$\omega^{ij} = c_k^{ij} w^k + f^{ij}, \tag{4.74}$$

where c_k^{ij} and f^{ij} are constant. From conditions (4.70) and (4.71) it follows that c_k^{ij} are structural constants of an N-dimensional Lie algebra

$$[e^i, e^j] = c_k^{ij} e^k, \tag{4.75}$$

and $f(e^i, e^j) = f^{ij}$ is a 2-cocycle on this Lie algebra, that is, the following relations are fulfilled:

$$f^{ij} = -f^{ji} \quad (\text{skew} - \text{symmetry}), \tag{4.76}$$

$$(df)(e^i, e^j, e^k) \equiv f([e^i, e^j], e^k) +$$
$$+ f([e^k, e^i], e^j) + f([e^j, e^k], e^i) =$$
$$= c_s^{ij} f^{sk} + c_s^{ki} f^{sj} + c_s^{jk} f^{si} = 0 \quad (\text{closedness}). \tag{4.77}$$

Moreover, the Lie algebra (4.75) is equipped with a non-degenerate bilinear form

$$\langle e^i, e^j \rangle = g^{ij1}$$

satisfying the relations

$$\langle e^i, e^j \rangle = \langle e^j, e^i \rangle, \tag{4.78}$$

$$\langle [e^i, e^j], e^k \rangle = \langle e^i, [e^j, e^k] \rangle. \tag{4.79}$$

Relation (4.79) is equivalent to relation (4.72):

$$g^{ks} c_s^{ij} + g^{is} c_s^{kj} = 0.$$

Conditions (4.78) and (4.79) mean that the Lie algebra (4.75) is equipped with the *invariant symmetric bilinear form* $\langle e^i, e^j \rangle$.

In the one-dimensional case there are no other conditions on the Lie algebra (4.75). We note that the condition of compatibility for an arbitrary *Lie–Poisson bivector* (4.74), that is, a Poisson bivector which is linear with respect to coordinates, with a constant Poisson bracket of hydrodynamic type, means precisely that the metric of this Poisson bracket is invariant on the Lie algebra (4.75). It is important to note that, as follows from Lemma 4.3, this fact is also valid in the case of a degenerate metric. Linear non-homogeneous Poisson structures of hydrodynamic type with degenerate metrics often arise in applications (in particular, see Examples 4.5 and 4.6 below).

Thus, any finite-dimensional Lie algebra equipped with a non-degenerate symmetric invariant scalar product and an arbitrary 2-cocycle on this Lie algebra defines a local non-homogeneous one-dimensional Poisson structure of hydrodynamic type, and consequently these Lie algebras completely describe Killing–Poisson bivectors on flat Riemannian or pseudo-Riemannian manifolds.

In particular, any semisimple Lie algebra defines a non-degenerate one-dimensional non-homogeneous local Poisson structure of hydrodynamic type. As the metric g^{ij1} we can take the Killing metric, which is non-degenerate in this case and satisfies conditions (4.78) and (4.79). The corresponding Poisson structures correspond to a class of Kac–Moody algebras (see [77]).

4.3.3   Kac–Moody algebras related to non-homogeneous Poisson brackets of hydrodynamic type

Let us consider the space S of sequences $\xi(x) = (\xi_1(x), \ldots, \xi_N(x))$, $\xi_i(x) \in C^\infty(S^1)$, and define the commutator on S:

$$[\xi(x), \eta(x)]_k = c_k^{ij} \xi_i(x) \eta_j(x), \tag{4.80}$$

where c_k^{ij} are the structural constants of an N-dimensional Lie algebra $\mathcal{G}$:

$$[e^i, e^j] = c_k^{ij} e^k.$$

The commutator (4.80) turns the space S into an infinite-dimensional Lie algebra, namely, a *loop algebra* $\widetilde{\mathcal{G}}_{S^1}$. For any semisimple Lie algebra $\mathcal{G}$ the Killing metric

$$\left\langle e^i, e^j \right\rangle_K = g^{ij}$$

defines a non-cohomologous to zero 2-cocycle h on the loop algebra $\widetilde{\mathcal{G}}_{S^1}$:

$$h(\xi(x), \eta(x)) = \int_{S^1} \langle \xi(x), \eta'(x) \rangle_K \, dx = \int_{S^1} g^{ij} \xi_i \eta_j' \, dx, \quad (4.81)$$

where $\eta' = d\eta/dx$.

This 2-cocycle $h(\xi, \eta)$ generates a non-trivial central extension $\widehat{\mathcal{G}}$ of the loop algebra $\widetilde{\mathcal{G}}_{S^1}$. All the extended algebras $\widehat{\mathcal{G}}$ obtained in this way (they are Kac–Moody algebras [77]) define an important class of non-homogeneous local Poisson structures of hydrodynamic type.

In fact, the operator M^{ij}, which is uniquely defined by the relation

$$\int_{S^1} \eta_i(x) (M^{ij} \xi_j(x)) \, dx =$$

$$= \int_{S^1} w^k(x) [\xi(x), \eta(x)]_k \, dx + h(\xi(x), \eta(x)), \quad (4.82)$$

is Hamiltonian.

One-dimensional non-homogeneous local Poisson structures of hydrodynamic type were studied by Dubrovin and Novikov [35] (see also [36], [37], [160]; in [35], [37], [160] there is an inaccuracy in the description of one-dimensional non-homogeneous local Poisson brackets of hydrodynamic type (4.67) which was corrected in [36]).

In the next section we shall consider a non-local generalization of these results.

Let us mention some examples of non-homogeneous systems of hydrodynamic type.

Example 4.5 *The 3-wave equations.*

The N-wave equations [162] are a Hamiltonian non-homogeneous system of hydrodynamic type and always possess a local Poisson structure of the form (4.67). In particular, for $N = 3$ the Poisson structure is generated by the simplest affine Kac–Moody algebra $\widehat{sl}(2)$, and the 3-wave interaction equations

$$\begin{cases} u_t^1 + v_1 u_x^1 = -2(v_2 - v_3)u^2 u^3, \\ u_t^2 + v_2 u_x^2 = 2(v_3 - v_1)u^1 u^3, \\ u_t^3 + v_3 u_x^3 = 2(v_1 - v_2)u^1 u^2, \end{cases} \qquad (4.83)$$

where v_i, $i = 1, 2, 3$, are constants, can be represented in the following Hamiltonian form:

$$u_t^i = M^{ij} \frac{\delta H}{\delta u^j(x)}, \qquad (4.84)$$

where H is the quadratic Hamiltonian

$$H = -\frac{1}{2} \int [v_1(u^1)^2 - v_2(u^2)^2 - v_3(u^3)^2]\, dx, \qquad (4.85)$$

and M^{ij} is the Hamiltonian operator

$$(M^{ij}) = \begin{pmatrix} 1 & 0 & 0 \\ 0 & -1 & 0 \\ 0 & 0 & -1 \end{pmatrix} \frac{d}{dx} +$$

$$+ \begin{pmatrix} 0 & -2u^3(x) & 2u^2(x) \\ 2u^3(x) & 0 & 2u^1(x) \\ -2u^2(x) & -2u^1(x) & 0 \end{pmatrix}. \qquad (4.86)$$

The Hamiltonian operator (4.86) is uniquely defined by (4.82) for the simple Lie algebra $sl(2)$ in a suitable basis $\{e^i\}$:

$$[e^1, e^2] = -2e^3, \qquad [e^1, e^3] = 2e^2, \qquad [e^2, e^3] = 2e^1, \qquad (4.87)$$

and the Killing metric in this basis has the form $\langle e^i, e^j \rangle_K = g^{ij}$,

$$(g^{ij}) = \begin{pmatrix} 1 & 0 & 0 \\ 0 & -1 & 0 \\ 0 & 0 & -1 \end{pmatrix}.$$

We shall need the 3-wave equations (4.83) in Section 5 for the study of equations of associativity in two-dimensional field theory.

In concrete examples Poisson structures of hydrodynamic type with degenerate metrics often arise, but their geometry has been insufficiently studied (see the general relations in Lemma 4.3). For example, a Poisson structure for the real reduction of the 2-wave interaction system

$$\begin{cases} u_t^1 = au^1u^2, \\ u_t^2 - au_x^2 = (u^1)^2, \end{cases} \tag{4.88}$$

is generated by the two-dimensional Lie algebra

$$[e^1, e^2] = e^1$$

and a degenerate 2-cocycle on its loop algebra:

$$(M^{ij}) = \begin{pmatrix} 0 & 0 \\ 0 & 1 \end{pmatrix} \frac{d}{dx} + \begin{pmatrix} 0 & u^1(x) \\ -u^1(x) & 0 \end{pmatrix},$$

and the Hamiltonian H is quadratic:

$$H = -\frac{1}{2} \int [(u^1)^2 - a(u^2)^2]\,dx.$$

Another important example with one degenerate and one non-degenerate metric is given by the KdV equation.

Example 4.6 *The Korteweg–de Vries equation.*

Let us regard the KdV equation as a non-homogeneous system of hydrodynamic type or, in this case, as an evolution system with respect to the space variable x:

$$\begin{cases} u_x^1 = u^2, \\ u_x^2 = u^3, \\ u_x^3 = -u_t^1 + 6u^1u^2. \end{cases} \tag{4.89}$$

In the paper [179] it is shown that the KdV system (4.89) is Hamiltonian with respect to some non-homogeneous Poisson structures of hydrodynamic type which are, in fact, induced by the well-known Magri and Gardner–Zakharov–Faddeev Poisson structures for the KdV equation. In the present author's paper [108] it is shown that the first Poisson structure of the KdV system (4.89) is generated by the simplest infinite-dimensional affine Kac–Moody algebra $\widehat{sl}(2)$, that is, it coincides with the Poisson structure of the 3-wave equations, and the second Poisson structure is defined by the unique three-dimensional nilpotent (non-Abelian) Lie algebra (this is the Lie algebra of type 2 in the Bianchi classification of all three-dimensional Lie algebras) and a 2-cocycle on its loop algebra.

In fact, the KdV system (4.89) has the following Hamiltonian representation:

$$u^i_x = \{u^i(t), H\}, \qquad H = -\frac{1}{2} \int [3(u^1(t))^2 - u^3(t)]\, dt, \qquad (4.90)$$

where the non-homogeneous Poisson structure of hydrodynamic type associated with the Magri bracket for the KdV equation (see [179]) is non-linear with respect to the field variables u^i:

$$\{u^i(t), u^j(t')\} = g^{ij}(u(t))\delta_t(t - t') +$$
$$+ b^{ij}_k(u(t))u^k_t \delta(t - t') + \omega^{ij}(u(t)), \qquad (4.91)$$

$$(g^{ij}) = \begin{pmatrix} 0 & 0 & 1 \\ 0 & -1 & 0 \\ 1 & 0 & 8u^1(t) \end{pmatrix}, \qquad (4.92)$$

$$(b^{ij}_k u^k_t) = \begin{pmatrix} 0 & 0 & 0 \\ 0 & 0 & 0 \\ 0 & 0 & 4u^1_t \end{pmatrix}, \qquad (4.93)$$

$$(\omega^{ij}) = \qquad (4.94)$$
$$= \begin{pmatrix} 0 & 2u^1(t) & 2u^2(t) \\ -2u^1(t) & 0 & -12(u^1(t))^2 + 2u^3(t) \\ -2u^2(t) & 12(u^1(t))^2 - 2u^3(t) & 0 \end{pmatrix}.$$

After a local quadratic unimodular change of variables, found by the present author in [108]:

$$\begin{cases} u^1 = (w^1 - w^3)/\sqrt{2}, \\ u^2 = w^2, \\ u^3 = (w^1 + w^3)/\sqrt{2} + (w^1 - w^3)^2, \end{cases} \qquad (4.95)$$

the Korteweg–de Vries system (4.89) can be rewritten in the form (see [108])

$$w^i_x = M^{ij}(t) \frac{\delta H}{\delta w^j(t)}, \qquad (4.96)$$

the Hamiltonian H being quadratic:

$$H = - \int [(w^1 - w^3)^2 - \sqrt{2}(w^1 + w^3)] \, dt, \qquad (4.97)$$

$$(M^{ij}(t)) = \begin{pmatrix} 1 & 0 & 0 \\ 0 & -1 & 0 \\ 0 & 0 & -1 \end{pmatrix} \frac{d}{dt} +$$

$$+ \begin{pmatrix} 0 & -2w^3(t) & 2w^2(t) \\ 2w^3(t) & 0 & 2w^1(t) \\ -2w^2(t) & -2w^1(t) & 0 \end{pmatrix}, \qquad (4.98)$$

where $M^{ij}(t)$ is the Poisson structure generated by the Kac–Moody algebra $\widehat{sl}(2)$.

The second non-homogeneous Poisson structure of hydrodynamic type for the KdV system (4.89) is compatible with the first one, but it is degenerate and has the form:

$$(M_2^{ij}(t)) = \frac{1}{2} \begin{pmatrix} 1 & 0 & 1 \\ 0 & 0 & 0 \\ 1 & 0 & 1 \end{pmatrix} \frac{d}{dt} +$$

$$+ (w^1 - w^3) \begin{pmatrix} 0 & 1 & 0 \\ -1 & 0 & -1 \\ 0 & 1 & 0 \end{pmatrix} +$$

$$+ \frac{1}{\sqrt{2}} \begin{pmatrix} 0 & 1 & 0 \\ -1 & 0 & 1 \\ 0 & -1 & 0 \end{pmatrix}, \qquad (4.99)$$

where the metric $g^{ij}(w)$ is degenerate. The Poisson structure (4.99) is associated with the Gardner–Zakharov–Faddeev Poisson bracket for the KdV equation and is generated by the three-dimensional nilpotent non-Abelian Lie algebra $\mathcal{G}_0$, a non-cohomologous to zero 2-cocycle and a degenerate invariant symmetric bilinear form on this Lie algebra. The Lie algebra $\mathcal{G}_0$ is the Lie algebra of type 2 in the Bianchi classification of three-dimensional Lie algebras (see [37]) and it is well known that any nilpotent non-Abelian Lie algebra always contains a subalgebra isomorphic to $\mathcal{G}_0$.

Correspondingly, the second Hamiltonian representation for the KdV system (4.89) has the form:

$$w^i_x = M^{ij}_2(t)\frac{\delta H}{\delta w^j(t)},$$

where H is a quadratic Hamiltonian:

$$H = -\frac{1}{2}\int[(w^1)^2 - (w^2)^2 - (w^3)^2]\,dt =$$

$$= -\frac{1}{2}\int\langle w, w\rangle_K\,dt. \qquad (4.100)$$

It is curious that the KdV system (4.89) has no other integrals of motion of hydrodynamic type except the Hamiltonians (4.97) and (4.100).

4.3.4   Reciprocal transformations and non-homogeneous systems of hydrodynamic type

Let us consider one more non-homogeneous system of hydrodynamic type which was integrated by Calogero in [14]. Recently, this system arose independently in the paper of Ganzha [61] in the study of non-homogeneous hydrodynamic type systems which have constant characteristic velocities and possess a higher first-order conservation law:

$$u^i_t = a^i u^i_x + u^i\sum_k(a^i - a^k)u^k, \qquad (4.101)$$

where $a^i \neq a^j$ for $i \neq j$, $i, j = 1, \ldots, N$.

Proposition 4.2 *By a reciprocal transformation and changes of variables the system (4.101) reduces to a homogeneous diagonal semi-Hamiltonian system of hydrodynamic type which can be integrated by the generalized hodograph method.*

First of all, we introduce the new field variables $w^i(x)$:

$$u^i = e^{w^i}.$$

Then the system (4.101) assumes the following form:

$$w^i_t = a^i w^i_x + \sum_k (a^i - a^k)e^{w^k}. \qquad (4.102)$$

Let us consider the so-called reciprocal transformations for the system (4.102) or, in other words, transformations of independent variables x and t of the form

$$\begin{cases} dx' = \varphi_1(x,t,w)\,dx - \psi_1(x,t,w)\,dt, \\ dt' = \varphi_2(x,t,w)\,dx - \psi_2(x,t,w)\,dt, \end{cases} \qquad (4.103)$$

where

$$\frac{\partial \varphi_i(x,t,w)}{\partial t} + \frac{\partial \psi_i(x,t,w)}{\partial x} = 0, \quad i = 1, 2, \qquad (4.104)$$

$$\Delta = \varphi_1 \psi_2 - \varphi_2 \psi_1 \neq 0.$$

Here $w^i(x,t)$ are arbitrary solutions of the system (4.102) and φ_i, ψ_i, $i = 1, 2$, are the densities of conservation laws and corresponding fluxes (4.104) of the system (4.102), respectively.

We consider the following two conservation laws of the system (4.102):

1. $\varphi_1 = -\sum_k e^{w^k}$, $\psi_1 = \sum_k a^k e^{w^k}$;

2. $\varphi_2 = 0$, $\psi_2 = 1$.

Using the relations

$$\begin{cases} w^i_x = w^i_{x'}\varphi_1 + w^i_{t'}\varphi_2, \\ w^i_t = -w^i_{x'}\psi_1 - w^i_{t'}\psi_2, \end{cases}$$

after the corresponding reciprocal transformation (4.103), (4.104) we obtain the following non-homogeneous system of hydrodynamic type:

$$w^i_{t'} = \left(\sum_k (a^i - a^k)e^{w^k} \right)(w^i_{x'} - 1). \qquad (4.105)$$

After elementary transformations

$$v^i = w^i - x'$$

and

$$x'' = -e^{-x'}$$

we obtain a diagonal semi-Hamiltonian homogeneous system of hydrodynamic type which is integrated by the generalized hodograph method according to Tsarev's Theorem 4.6 ([178], [181]):

$$v_{t'}^i = \left(\sum_k (a^i - a^k) e^{v^k} \right) v_{x''}^i. \tag{4.106}$$

4.4 Killing–Poisson bivectors on spaces of constant Riemannian curvature and bi-Hamiltonian structure of the generalized Heisenberg ferromagnets

4.4.1 Non-local non-homogeneous Poisson brackets of hydrodynamic type

Let us consider now *non-local non-homogeneous one-dimensional Poisson structures of hydrodynamic type*

$$\{u^i(x), u^j(y)\} = g^{ij}(u(x))\delta_x(x-y) + b_k^{ij}(u(x))u_x^k\delta(x-y) +$$

$$+K u_x^i \left(\frac{d}{dx} \right)^{-1} u_x^j \delta(x-y) + \omega^{ij}(u(x))\delta(x-y). \tag{4.107}$$

For any Hamiltonian of hydrodynamic type, Poisson structures of the form (4.107) also always generate non-homogeneous systems of hydrodynamic type.

Lemma 4.5 *The expression (4.107) defines a Poisson structure if and only if it is the sum of two compatible Poisson structures, namely, a non-local Poisson bracket of hydrodynamic type (4.45) and a classic Poisson bivector $\omega^{ij}(u)$ on a manifold, that is, an ultralocal Poisson bracket*

$$\{u^i(x), u^j(y)\} = \omega^{ij}(u(x))\delta(x-y). \tag{4.108}$$

The conditions of compatibility for these two Poisson brackets have the form

$$\Phi^{ijk,\alpha} = \Phi^{kij,\alpha}, \tag{4.109}$$

where

$$\Phi^{ijk} = g^{is}\frac{\partial\omega^{jk}}{\partial u^s} - b_s^{ij}\omega^{sk} - b_s^{ik}\omega^{js},$$

$$\frac{\partial\Phi^{ijk}}{\partial u^r} = \sum_{(i,j,k)}\left[b_r^{si}\frac{\partial\omega^{jk}}{\partial u^s} + \right.$$

$$\left. + \left(\frac{\partial b_r^{ij}}{\partial u^s} - \frac{\partial b_s^{ij}}{\partial u^r}\right)\omega^{sk} + K\omega^{ij}\delta_r^k\right]. \tag{4.110}$$

In Lemma 4.5 non-degeneracy of the metric is not assumed.

Lemma 4.6 *If the metric g^{ij} in (4.107) is non-degenerate, then the conditions of compatibility (4.109) and (4.110) can be rewritten in the form:*

$$\nabla^i\omega^{jk} + \nabla^j\omega^{ik} = 0, \tag{4.111}$$

$$\nabla_r\nabla^i\omega^{jk} = \sum_{(i,j,k)}(-K\omega^{ij}\delta_r^k), \tag{4.112}$$

where ∇^i is the covariant derivative generated by the metric g^{ij}.

Theorem 4.11 *The Poisson structures (4.45) and (4.108) are compatible if and only if $\omega^{ij}(u)$ is a Killing–Poisson bivector on a manifold (M, g^{ij}) of constant Riemannian curvature K.*

Theorem 4.12 *For $N = 2$ the tensor $\omega^{ij}(u)$ is a Killing–Poisson bivector on the manifold (M, g_{ij}) of constant Riemannian curvature K if and only if in the canonical coordinates (u^1, u^2) (see (4.52))*

$$(\omega^{ij}(u)) = c\lambda^2(u)\begin{pmatrix} 0 & 1 \\ -1 & 0 \end{pmatrix}, \tag{4.113}$$

where c is an arbitrary constant.

Each Killing–Poisson bivector $\omega^{ij}(u)$ on the manifold (M, g_{ij}) of constant Riemannian curvature K defines a pair of compatible Poisson structures

$$M_1^{ij} = g^{ij}(u)\frac{d}{dx} - g^{is}(u)\Gamma_{sk}^j(u)u_x^k + Ku_x^i\left(\frac{d}{dx}\right)^{-1}u_x^j, \tag{4.114}$$

$$M_2^{ij} = \omega^{ij}(u), \tag{4.115}$$

generating the integrable hierarchy of N-component Heisenberg ferromagnets

$$S_t = [S, S_{xx}], \qquad S^2 = 1, \tag{4.116}$$

where $S = (S^1, \ldots, S^{N+1})$, and $[\cdot, \cdot]$ is the commutator in a corresponding $(N+1)$-dimensional Lie algebra.

4.4.2 The Heisenberg ferromagnet and non-local Poisson structure of hydrodynamic type

The classic Heisenberg ferromagnet corresponds to the case of the two-dimensional sphere $(N = 2)$. For the sphere

$$S^{1^2} + S^{2^2} + S^{3^2} = 1$$

in the coordinates of stereographic projection

$$S^1 = \frac{u^1}{P}, \qquad S^2 = \frac{u^2}{P}, \qquad S^3 = \frac{P-1}{P},$$

where

$$P = \frac{(u^1)^2 + (u^2)^2 + 1}{2},$$

the metric has the form

$$(g_{ij}) = \frac{1}{P^2} \begin{pmatrix} 1 & 0 \\ 0 & 1 \end{pmatrix}.$$

The non-local Poisson structure (4.114) generated by the metric has the form

$$M_1 = P^2 \begin{pmatrix} 1 & 0 \\ 0 & 1 \end{pmatrix} \frac{d}{dx} + P \begin{pmatrix} u^1 u_x^1 + u^2 u_x^2 & u^1 u_x^2 - u^2 u_x^1 \\ u^2 u_x^1 - u^1 u_x^2 & u^1 u_x^1 + u^2 u_x^2 \end{pmatrix} +$$

$$+ \begin{pmatrix} u_x^1 \\ u_x^2 \end{pmatrix} \left(\frac{d}{dx} \right)^{-1} \circ \begin{pmatrix} u_x^1 & u_x^2 \end{pmatrix}. \tag{4.117}$$

According to Theorem 4.12 the unique (up to a constant factor) Killing–Poisson bivector on the two-dimensional sphere has the following form in these coordinates:

$$M_2 = (\omega^{ij}(u)) = \begin{pmatrix} 0 & -P^2 \\ P^2 & 0 \end{pmatrix}. \qquad (4.118)$$

Now, following the usual bi-Hamiltonian scheme and applying the recursion operator

$$R = M_1(M_2)^{-1}$$

to the system of translations with respect to x, we obtain the system

$$\begin{pmatrix} u^1 \\ u^2 \end{pmatrix}_t = R \begin{pmatrix} u^1 \\ u^2 \end{pmatrix}_x,$$

or

$$\begin{cases} u_t^1 = u_{xx}^2 + \left(u^2(u_x^1)^2 - 2u^1 u_x^1 u_x^2 - u^2(u_x^2)^2 \right)/P, \\ u_t^2 = -u_{xx}^1 - \left(u^1(u_x^2)^2 - 2u^2 u_x^1 u_x^2 - u^1(u_x^1)^2 \right)/P, \end{cases} \qquad (4.119)$$

which coincides with the classic equations of a Heisenberg ferromagnet

$$\vec{S}_t = \vec{S} \times \vec{S}_{xx}, \qquad \vec{S}^2 = 1.$$

The corresponding bi-Hamiltonian representation has the form

$$\begin{pmatrix} u^1 \\ u^2 \end{pmatrix}_t = M_1 \begin{pmatrix} \delta G/\delta u^1(x) \\ \delta G/\delta u^2(x) \end{pmatrix} = M_2 \begin{pmatrix} \delta H/\delta u^1(x) \\ \delta H/\delta u^2(x) \end{pmatrix}, \qquad (4.120)$$

where

$$G = \int \frac{u^2 u_x^1 - u^1 u_x^2}{(2P-1)P} \, dx, \qquad H = \frac{1}{2} \int \frac{(u_x^1)^2 + (u_x^2)^2}{P^2} \, dx. \qquad (4.121)$$

4.4.3   Killing–Poisson bivectors on spaces of constant Riemannian curvature

The bivector (4.118) is a result of the restriction of the Lie–Poisson bivector

$$\Omega = \begin{pmatrix} 0 & S^3 & -S^2 \\ -S^3 & 0 & S^1 \\ S^2 & -S^1 & 0 \end{pmatrix}, \qquad (4.122)$$

defined in Euclidean space with coordinates S^1, S^2, S^3, to the two-dimensional sphere.

In the paper [152] the present author and Ferapontov found a general description for all Killing–Poisson bivectors on spaces of constant Riemannian curvature. Here we mention these results for an N-dimensional sphere. Let c_k^{ij} be structural constants of some $(N+1)$-dimensional Lie algebra

$$[e^i, e^j] = c_k^{ij} e^k, \quad i, j, k = 1, \ldots, N+1,$$

such that the following relations are satisfied:

$$c_k^{ij} + c_i^{kj} = 0, \qquad (4.123)$$

that is, this Lie algebra is equipped with an invariant scalar product

$$\langle [e^i, e^j], e^k \rangle = \langle e^i, [e^j, e^k] \rangle,$$

which has the standard Euclidean form in the chosen coordinate system:

$$\langle e^i, e^j \rangle = \delta^{ij}.$$

We consider the skew-symmetric Lie–Poisson bivector

$$\Omega^{ij} = c_k^{ij} S^k$$

and restrict it to the N-dimensional sphere

$$\sum_{k=1}^{N+1} (S^k)^2 = 1.$$

Because of the relations (4.123) this restriction is possible. We shall present the explicit form of the bivector ω^{ab}, $a, b = 1, \ldots, N$, obtained by restricting the Lie–Poisson bivector $\Omega^{ij} = c_k^{ij} S^k$ to the N-dimensional sphere in the coordinates $u^1, \ldots, u^N$ of stereographic projection

$$S^1 = \frac{u^1}{P}, \ldots, S^N = \frac{u^N}{P}, \quad S^{N+1} = \frac{P-1}{P}, \qquad (4.124)$$

$$P = \frac{1}{2} \left(\sum_{a=1}^{N} (u^a)^2 + 1 \right),$$

$$\omega^{ab} = P \sum_{s=1}^{N} (c_{N+1}^{sa} u^s u^b - c_{N+1}^{sb} u^s u^a +$$

$$+ c_s^{ab} u^s) + c_{N+1}^{ab} (P - 1) P. \tag{4.125}$$

The bivector ω^{ab} is a Killing–Poisson bivector on the N-dimensional sphere. This follows from the fact that Ω^{ij} is a Killing–Poisson bivector of the enveloping Euclidean space equipped with the corresponding flat connection. Moreover, any Killing–Poisson bivector on an N-dimensional sphere can be obtained by the described construction from a certain Lie algebra equipped with an invariant Euclidean scalar product [152]. In particular, the Killing–Poisson bivector (4.118) on a two-dimensional sphere is connected with the Lie algebra $so(3)$.

Compatible Poisson structures defined by Killing–Poisson bivectors on an N-dimensional sphere generate the hierarchy of generalized Heisenberg ferromagnets (4.116) for the corresponding $(N + 1)$-dimensional Lie algebra equipped with an invariant Euclidean scalar product.

4.5 Homogeneous Poisson structures of differential-geometric type

4.5.1 General Homogeneous Poisson brackets of arbitrary orders

In [35] Dubrovin and Novikov introduced homogeneous differential-geometric Poisson brackets of arbitrary order n as a natural generalization of Poisson structures of hydrodynamic type and posed the problem of classifying these brackets. In the one-dimensional case homogeneous non-degenerate Poisson brackets of differential-geometric type have the form:

$$\{u^i(x), u^j(y)\} = g^{ij}(u(x))\delta_{(n)}(x - y) +$$

$$+ b_k^{ij}(u(x))u_x^k \delta_{(n-1)}(x - y) +$$

$$+ [c_k^{ij}(u(x))u_{xx}^k + c_{kl}^{ij}(u(x))u_x^k u_x^l]\delta_{(n-2)}(x - y) + \cdots +$$

$$+ [d_k^{ij}(u(x))u_{(n)}^k + \cdots]\delta(x - y), \quad \det(g^{ij}(u)) \neq 0, \tag{4.126}$$

where each term has degree of homogeneity n with respect to the natural grading

$$\deg(hg) = \deg h + \deg g, \qquad \deg f(u(x)) = \deg u(x) = 0,$$

$$\deg \frac{d^k u}{dx^k} = \deg \delta_{(k)}(x - y) = k, \qquad f_{(k)} = \frac{d^k f}{dx^k}.$$

For $n = 0$ we obtain the classic finite-dimensional Poisson structures $g^{ij}(u)$ on a manifold M and the case $n = 1$ corresponds to homogeneous Poisson structures of hydrodynamic type.

The Dubrovin–Novikov structures (4.126) define different curious geometries on manifolds M with local coordinates $u^1, \ldots, u^N$: under local changes of coordinates the coefficient $g^{ij}(u)$ is transformed as a metric on M (which is symmetric for any odd n and skew-symmetric for any even n), and the coefficients

$$g_{is}(u)b_j^{sk}(u), \; g_{is}(u)c_j^{sk}(u), \ldots, g_{is}(u)d_j^{sk}(u)$$

are transformed like Christoffel symbols of affine connections on M. At present, a complete classification of homogeneous Poisson brackets of differential-geometric type (4.126) has been obtained only in the following cases: $n = 0$ (Darboux), $n = 1$ (Dubrovin and Novikov, [34]), $n = 2$ (Potemin, [172]; Doyle, [26]). In [172] and [26] the case $n = 3$ is partially investigated but a complete classification is not obtained. Moreover, no system of equations was found that arises in applications and possesses such homogeneous Poisson structures of order $n > 1$. Later the first examples of such homogeneous Poisson structures for some important non-linear systems (such as the equations of associativity and the Monge–Ampère equations) were found in [50], [154].

In [160] Novikov formulated the following conjecture: for any $n > 0$ the last connection $\widetilde{\Gamma}_{jk}^i = g_{js}(u)d_k^{si}(u)$ in (4.126) is symmetric and its curvature tensor vanishes, that is, the connection is flat. This conjecture of Novikov was proved in [172], [26] for arbitrary n.

We present briefly here the results of the study of general Dubrovin–Novikov structures.

4.5.2   Homogeneous Poisson brackets of the second order

Let us consider homogeneous differential-geometric Poisson brackets of the second order:

$$
\begin{aligned}
\{u^i(x), u^j(y)\} &= g^{ij}(u(x))\delta''(x-y) + \\
&+ b_k^{ij}(u(x))u_x^k\delta'(x-y) + [c_k^{ij}(u(x))u_{xx}^k + \\
&+ c_{kl}^{ij}(u(x))u_x^k u_x^l]\delta(x-y), \quad \det(g^{ij}(u)) \neq 0. \quad (4.127)
\end{aligned}
$$

We introduce the coefficients $\Gamma_{jk}^i(u)$ and $\widetilde{\Gamma}_{jk}^i(u)$ by the following relations:

$$
b_k^{ij} = -2g^{is}\Gamma_{sk}^j, \qquad c_k^{ij} = -g^{is}\widetilde{\Gamma}_{sk}^j. \quad (4.128)
$$

The connection $\Gamma_{jk}^i(u)$ must be compatible with the skew-symmetric metric $g^{ij}(u)$, that is, we have a symplectic connection $\Gamma_{jk}^i(u)$ on the almost symplectic manifold (M, g_{ij}) with the almost symplectic structure g_{ij}, and the connection $\widetilde{\Gamma}_{jk}^i(u)$ must be flat, with

$$
\widetilde{\Gamma}_{jk}^i(u) = \frac{1}{2}(\Gamma_{jk}^i(u) + \Gamma_{kj}^i(u)).
$$

In *special coordinates*, in which $\widetilde{\Gamma}_{jk}^i(u) = 0$ (flat coordinates of the connection $\widetilde{\Gamma}_{jk}^i(u)$), the relation $c_{kl}^{ij}(u) = 0$ is also always satisfied, that is, in these coordinates the Poisson bracket has the form:

$$
\begin{aligned}
\{u^i(x), u^j(y)\} &= g^{ij}(u(x))\delta''(x-y) + \\
&+ b_k^{ij}(u(x))u_x^k\delta'(x-y), \quad \det(g^{ij}(u)) \neq 0. \quad (4.129)
\end{aligned}
$$

Proposition 4.3 ([172]) *The expression (4.129) defines a Poisson bracket if and only if the relations*

$$
g^{ij} = -g^{ji}, \quad (4.130)
$$

$$
\frac{\partial g^{ij}}{\partial u^k} = b_k^{ij}, \quad (4.131)
$$

$$
b_s^{ij}g^{sk} = b_s^{jk}g^{si} \quad (4.132)
$$

are satisfied.

We can show that in this case the torsion tensor

$$T_{ijk}(u) = g_{is}T^s_{jk}, \quad T^j_{sk} = \Gamma^j_{sk} - \Gamma^j_{ks},$$

is absolutely skew-symmetric.

Theorem 4.13 ([172], [26]) *In special coordinates we have*

$$g_{ij}(u) = T_{ijk}u^k + g^0_{ij}, \tag{4.133}$$

where $T_{ijk} = $ const, $g^0_{ij} = $ const, *and the Poisson bracket (4.129) has the form:*

$$\{u^i(x), u^j(y)\} = \frac{d}{dx}g^{ij}(u)\frac{d}{dx}\delta(x - y). \tag{4.134}$$

The arbitrary constants $T_{ijk} = $ const, $g^0_{ij} = $ const, *which are skew-symmetric with respect to any pair of indices, define the Poisson bracket according to formulae (4.133), (4.134).*

Homogeneous differential-geometric Poisson structures of the second order reduce to a partial class of local symplectic structures of zero order, which were completely studied in Section 2.1.3. In fact, after the transformation $u^i = v^i_x$ the Poisson bracket (4.134) becomes a local Poisson bracket of zero order:

$$\{v^i(x), v^j(y)\} = -g^{ij}(v_x)\delta(x - y),$$

where

$$g_{ij}(v_x) = T_{ijk}v^k_x + g^0_{ij}, \quad T_{ijk} = \text{const}, \quad g^0_{ij} = \text{const},$$

is a symplectic structure of the form (2.21).

We introduce the 2-form $\Omega = g_{ij}(u)\, du^i \wedge du^j$.

Theorem 4.14 ([173]) *A homogeneous Poisson bracket of the second order (4.127) reduces to the constant form* $g^{ij}_0 \delta''(x - y)$ *if and only if* $d\Omega = 0$ *(or, what is equivalent, if the torsion tensor vanishes* $T^i_{jk}(u) = 0$*).*

We note that

$$(dg)_{ijk} = 3T_{ijk}.$$

4.5.3   Homogeneous Poisson brackets of the third order

Now let us consider the homogeneous Poisson brackets of the third order:

$$\{u^i(x), u^j(y)\} = g^{ij}(u(x))\delta'''(x-y) + b_k^{ij}(u(x))u_x^k\delta''(x-y) +$$
$$+[c_k^{ij}(u(x))u_{xx}^k + c_{kl}^{ij}(u(x))u_x^k u_x^l]\delta'(x-y) +$$
$$+[d_k^{ij}(u(x))u_{xxx}^k + d_{kl}^{ij}(u(x))u_{xx}^k u_x^l +$$
$$+d_{klm}^{ij}(u(x))u_x^k u_x^l u_x^m]\delta(x-y), \quad \det(g^{ij}(u)) \neq 0. \qquad (4.135)$$

Since the last connection in (4.135) $\tilde{\Gamma}_{sk}^j(u) = g_{si}(u)d_k^{ij}(u)$ is flat [172], [26], it follows that there exist special local coordinates on M in which $\tilde{\Gamma}_{sk}^j = 0$.

In the special coordinates the relations

$$d_{lm}^{jk}(u) = 0, \qquad d_{lmn}^{jk}(u) = 0 \qquad (4.136)$$

are always satisfied.

Conjecture. Apparently, for any $n > 0$ there exist local coordinates in which the coefficient of $\delta(x-y)$ in the homogeneous differential-geometric Poisson bracket (4.126) of order n vanishes; at least this is true for $n = 1, 2, 3$.

In the special coordinates, the Poisson bracket of the third order (4.135) has the form

$$\{u^i(x), u^j(y)\} = g^{ij}(u(x))\delta'''(x-y) +$$
$$+b_k^{ij}(u(x))u_x^k\delta''(x-y) + [c_k^{ij}(u(x))u_{xx}^k +$$
$$+c_{kl}^{ij}(u(x))u_x^k u_x^l]\delta'(x-y), \quad \det(g^{ij}(u)) \neq 0. \quad (4.137)$$

Theorem 4.15 ([172]) *The expression (4.137) defines a Poisson bracket if and only if the relations (4.138)–(4.144) are satisfied:*

$$\frac{\partial g^{ij}}{\partial u^k} = c_k^{ij} + c_k^{ji}, \qquad (4.138)$$

$$b_k^{ij} = 2c_k^{ij} + c_k^{ji}, \qquad (4.139)$$

$$c_l^{ij} g^{lk} = -c_l^{kj} g^{li}, \qquad (4.140)$$

$$c_l^{ij}g^{lk} + c_l^{jk}g^{li} + c_l^{ki}g^{lj} = 0, \tag{4.141}$$

$$2c_{lm}^{ij} = \frac{\partial c_l^{ij}}{\partial u^m} + \frac{\partial c_m^{ij}}{\partial u^l}, \tag{4.142}$$

$$g^{kl}\frac{\partial c_l^{ij}}{\partial u^m} = c_l^{ik}c_m^{lj} - c_l^{ki}c_m^{lj} - c_l^{kj}\frac{\partial g^{li}}{\partial u^m}, \tag{4.143}$$

$$\sum_{(m,n,p)} A_{rpij}c_m^{ki}c_n^{lj} = 0, \tag{4.144}$$

where

$$A_{rpij} = -\left(\frac{\partial c_{rjp}}{\partial u^i} + \frac{\partial c_{jip}}{\partial u^r} + \frac{\partial c_{rij}}{\partial u^p}\right), \tag{4.145}$$

and $\sum_{(m,n,p)}$ *means summation over all cyclic permutations of the elements* (m, n, p), $c_{mnk} = g_{mj}g_{ni}c_k^{ij}$, $g^{ij}g_{jk} = \delta_k^i$.

The conditions (4.138)–(4.144) are considerably simpler for differential-geometric objects with lower indices:

$$\frac{\partial g_{mn}}{\partial u^k} = -c_{mnk} - c_{nmk}, \tag{4.146}$$

$$c_{mnk} = -c_{mkn}, \tag{4.147}$$

$$c_{mnk} + c_{nkm} + c_{kmn} = 0, \tag{4.148}$$

$$\frac{\partial c_{mnk}}{\partial u^l} = -c_{pml}g^{pq}c_{qnk}, \tag{4.149}$$

$$\sum_{(m,n,p)} A_{rp}^{lk}c_{qlm}c_{skn} = 0, \tag{4.150}$$

where

$$A_{rp}^{lk} = -g^{li}g^{kj}\left(\frac{\partial c_{rjp}}{\partial u^i} + \frac{\partial c_{jip}}{\partial u^r} + \frac{\partial c_{rij}}{\partial u^p}\right). \tag{4.151}$$

Theorem 4.16 ([172], [26]) *In the special coordinates*

$$\frac{\partial^2 c_{mnk}}{\partial u^l \partial u^p} = 0, \tag{4.152}$$

that is,

$$c_{mnk} = c_{mnkl}u^l + c_{mnk}^0, \tag{4.153}$$

where c_{mnkl} and c^0_{mnk} are constants and the metric with lower indices is quadratic:

$$g_{mn}(u) = g_{mnpq}u^p u^q + g_{mnp}u^p + g^0_{mn}, \qquad (4.154)$$

where g_{mnpq}, g_{mnp} and g^0_{mn} are constants.

In the special coordinates the values $c_{mnp}(u)$ coincide, up to a constant factor, with the values of the torsion tensor with lower indices for the connection

$$c^{ij}_k = -3g^{is}\Gamma^j_{sk}$$

which is compatible with the metric g_{ij}. The torsion tensor c_{mnp} satisfying relations (4.146)–(4.150) and the constant matrix g^0_{mn} completely define the non-degenerate homogeneous differential-geometric Poisson bracket of the third order (4.135). The complete classification of these Poisson brackets has not yet been found. It would be interesting to elucidate the algebraic nature of the structural constants $c_{mnkl}, c^0_{mnk}, g^0_{mn}$ which define non-degenerate homogeneous differential-geometric Poisson brackets of the third order. Possibly they are connected with well-known algebraic structures, but this problem has not yet been solved. We note that in the papers [172], [173], [26] the following curious fact was not noticed.

Proposition 4.4 *In the special coordinates any homogeneous Poisson structure of the third order (4.137) has the form:*

$$M^{ij} = \frac{d}{dx} \circ \left(g^{ij}(u)\frac{d}{dx} + c^{ij}_k(u)u^k_x \right) \circ \frac{d}{dx}.$$

The proof follows from relations (4.138), (4.139), and (4.142).

Correspondingly, after the transformation $u^i = v^i_x$ the Poisson structure (4.137) is transformed to a Poisson structure of the first order. Namely:

$$\widetilde{M}^{ij} = -g^{ij}(v_x)\frac{d}{dx} - c^{ij}_k(v_x)v^k_{xx}.$$

Moreover, it is necessary to describe all first-order Poisson structures of the form

$$M^{ij} = g^{ij}(v_{(n)})\frac{d}{dx} - c^{ij}_k(v_{(n)})v^k_{(n+1)}.$$

After the transformation $u^i = v^i_{(n)}$ all these Poisson structures become homogeneous Poisson structures of order $2n + 1$. In this connection the conjecture of the present author is the following: all non-degenerate homogeneous Poisson structures reduce in this way either to Poisson structures of zero order (for structures of even order) or to Poisson structures of the first order (for structures of odd order). At least for $n = 0, 1, 2, 3$ this is true. If the conjecture is true, then from the description of all non-degenerate zero-order Poisson structures, which was obtained in Section 2.1.4, it follows that all homogeneous Poisson structures of even order (not equal to 2) can be reduced to a constant form by a local change of coordinates on the manifold.

Theorem 4.17 ([172], [26]) *A non-degenerate homogeneous differential-geometric Poisson bracket (4.135) of the third order can be reduced by local changes of coordinates on the manifold M to the constant form*

$$\{u^i(x), u^j(y)\} = g_0^{ij} \delta'''(x - y), \tag{4.155}$$

where $g_0^{ij} = \mathrm{const}$, if and only if the connection

$$c_k^{ij}(u) = -3g^{is}(u)\Gamma^j_{sk}(u)$$

has a zero torsion tensor, that is, it is symmetric (this is equivalent to the condition that the last connection $\tilde{\Gamma}^j_{sk}(u) = g_{si}(u)d_k^{ij}(u)$ is compatible with the metric $g^{ij}(u)$).

142 O.I. MOKHOV

5 Equations of associativity in two-dimensional topological field theory and non-diagonalizable integrable systems of hydrodynamic type

5.1 Equations of associativity as non-diagonalizable integrable homogeneous systems of hydrodynamic type

Let us consider a function of N variables $F(t^1, \ldots, t^N)$ such that the following two conditions are satisfied.

1. *The matrix*

$$\eta_{\alpha\beta} = \frac{\partial^3 F}{\partial t^1 \partial t^\alpha \partial t^\beta} \qquad (\alpha, \beta = 1, \ldots, N)$$

   is constant and non-degenerate.

   Note that the matrix $\eta_{\alpha\beta}$ defines the dependence of the function F on the fixed variable t^1 (up to quadratic polynomials in the variables $t^1, \ldots, t^N$).

2. *For any $t = (t^1, \ldots, t^N)$ the functions*

$$c^\alpha_{\beta\gamma}(t) = \eta^{\alpha\mu} \frac{\partial^3 F}{\partial t^\mu \partial t^\beta \partial t^\gamma} \qquad (here \quad \eta^{\alpha\mu}\eta_{\mu\beta} = \delta^\alpha_\beta)$$

   define the structure of an associative algebra $A(t)$ in N-dimensional linear space with basis $e_1, \ldots, e_N$ and multiplication

$$e_\beta \circ e_\gamma = c^\alpha_{\beta\gamma}(t)e_\alpha.$$

The condition of associativity

$$(e_\alpha \circ e_\beta) \circ e_\gamma = e_\alpha \circ (e_\beta \circ e_\gamma),$$

which is equivalent to the relation

$$c^\mu_{\alpha\beta}c^\nu_{\mu\gamma} = c^\nu_{\alpha\mu}c^\mu_{\beta\gamma}$$

for the structural constants of the algebra $A(t)$, results in a complicated overdetermined system of non-linear third-order partial differential equations for the function F. This system is known in two-dimensional topological field theory as the *equations of associativity* or the *Witten–Dijkgraaf–H. Verlinde–E. Verlinde–Dubrovin system* (see [189], [190], [17], [30]; all necessary physical motivations and the theory of integrability of the equations of associativity can be found in the survey of Dubrovin [29]):

$$\eta^{\mu\lambda} \frac{\partial^3 F}{\partial t^\lambda \partial t^\alpha \partial t^\beta} \frac{\partial^3 F}{\partial t^\nu \partial t^\mu \partial t^\gamma} = \eta^{\mu\lambda} \frac{\partial^3 F}{\partial t^\nu \partial t^\alpha \partial t^\mu} \frac{\partial^3 F}{\partial t^\lambda \partial t^\beta \partial t^\gamma}. \quad (5.1)$$

Note that the equations of associativity (5.1) also play one of the key roles in the theory of Gromov–Witten invariants, which is being developed at present, the theory of quantum cohomology, and some classic problems of algebraic geometry, in particular, in the problem on the number $n(d)$ of rational curves of degree d on the projective plane $\mathbf{P}^2$ that pass through $3d - 1$ generic points or, in the more general case, on the number $n(d; k_2, \ldots, k_{r+1})$ of rational curves of degree d in projective space $\mathbf{P}^{r+1}$ that cross k_s hyperplanes of codimension s such that

$$\sum_{s=2}^{r+1} k_s(s - 1) = (r + 2)d + r - 2$$

(Ruan and Tian [176], Kontsevich and Manin [79], Kontsevich [78]).

The variable t^1 in the equations of associativity (5.1) is an initially fixed variable and, according to condition 1, the assignment of an arbitrary constant non-degenerate symmetric matrix $\eta_{\alpha\beta}$ for the function F simply determines the dependence of this function on the variable t^1 (by virtue of its definition the function F is always considered up to quadratic polynomials in the variables $t^1, \ldots, t^N$):

$$F = \frac{1}{6}\eta_{11}(t^1)^3 + \sum_{\beta=2}^{N} \frac{1}{2}\eta_{1\beta}(t^1)^2 t^\beta +$$

$$+ \sum_{\alpha \geq 2} \sum_{\beta > \alpha} \eta_{\alpha\beta} t^1 t^\alpha t^\beta + \sum_{\alpha \geq 2} \frac{1}{2}\eta_{\alpha\alpha} t^1 (t^\alpha)^2 + f(t^2, \ldots, t^N).$$

Correspondingly, for the given metric $\eta_{\alpha\beta}$ the equations of associativity are equivalent to a system of non-linear partial differential equations for the function $f(t^2, \ldots, t^N)$.

From condition 2 it follows immediately that multiplication in the algebra $A(t)$ is commutative, and from condition 1 it follows that the element e_1 is always a unit in the algebra $A(t)$:

$$e_1 \circ e_\mu = c^\nu_{1\mu} e_\nu = \eta^{\nu\lambda} \eta_{\lambda\mu} e_\nu = e_\mu.$$

Thus, for any t the algebra $A(t)$ is a commutative associative algebra with a unit. Moreover, the algebra $A(t)$ is equipped with a non-degenerate symmetric bilinear form

$$\langle e_\alpha, e_\beta \rangle = \eta_{\alpha\beta}$$

that is *invariant* with respect to multiplication in the algebra, that is,

$$\langle e_\alpha \circ e_\beta, e_\gamma \rangle = \langle e_\alpha, e_\beta \circ e_\gamma \rangle.$$

A finite-dimensional commutative associative algebra possessing a unit and equipped with a non-degenerate invariant symmetric bilinear form is called a *Frobenius algebra*. Consequently, the equations of associativity (5.1) describe N-parametric deformations of N-dimensional Frobenius algebras.

The first equations of associativity arise for $N = 3$. For $N = 3$ the following two essentially different types of dependence of the function F on the fixed variable t^1 were considered by Dubrovin in [29]:

$$F = \frac{1}{2}(t^1)^2 t^3 + \frac{1}{2} t^1 (t^2)^2 + f(t^2, t^3),$$

which corresponds to the antidiagonal metric

$$(\eta_{\alpha\beta}) = \begin{pmatrix} 0 & 0 & 1 \\ 0 & 1 & 0 \\ 1 & 0 & 0 \end{pmatrix},$$

and

$$F = \frac{1}{6}(t^1)^3 + t^1 t^2 t^3 + f(t^2, t^3),$$

which corresponds to the metric with the same signature but of the form

$$(\eta_{\alpha\beta}) = \begin{pmatrix} 1 & 0 & 0 \\ 0 & 0 & 1 \\ 0 & 1 & 0 \end{pmatrix}.$$

The corresponding equations of associativity reduce to the following two non-linear equations of the third order for the function of two independent variables $f = f(x, t)$ (henceforth, everywhere we use the notation $x = t^2$, $t = t^3$):

$$f_{ttt} = f_{xxt}^2 - f_{xxx}f_{xtt}, \tag{5.2}$$

$$f_{xxx}f_{ttt} - f_{xxt}f_{xtt} = 1, \tag{5.3}$$

respectively. Equation (5.2), which will be the main object under investigation in Section 5, describes quantum cohomology, namely, the deformations of the cohomology ring of the projective plane $\mathbf{P}^2$. The solution of the problem on the number $n(d)$ of rational curves of degree d on $\mathbf{P}^2$ which pass through $3d - 1$ generic points, as Kontsevich showed, is also connected with equation (5.2). Let us consider the series

$$\phi(x, t) = \sum_{d=1}^{\infty} n(d) \frac{t^{3d-1}}{(3d-1)!} e^{dx}.$$

This series must satisfy the associativity equation (5.2), whence Kontsevich's recursion relations for the numbers $n(d)$ are obtained (see [78], [79], [176]): $n(1) = 1$, and for $d \geq 2$

$$n(d) = \sum_{\substack{k, l \geq 1, \\ k + l = d}} n(k)n(l)k^2 l \left[l \left(\begin{array}{c} 3d - 4 \\ 3k - 2 \end{array} \right) - k \left(\begin{array}{c} 3d - 4 \\ 3k - 1 \end{array} \right) \right].$$

Generalizations of this problem to projective spaces of arbitrary dimensions and other varieties are also always connected with the equations of associativity describing the deformations of cohomology rings of the corresponding manifolds [78], [79], [176].

Theorem 5.1 ([124], [126]) *Equations (5.2) and (5.3) are equivalent to integrable non-diagonalizable homogeneous systems of hydrodynamic type*

$$\left(\begin{array}{c} a \\ b \\ c \end{array} \right)_t = \left(\begin{array}{ccc} 0 & 1 & 0 \\ 0 & 0 & 1 \\ -c & 2b & -a \end{array} \right) \left(\begin{array}{c} a \\ b \\ c \end{array} \right)_x, \tag{5.4}$$

$$\begin{pmatrix} a \\ b \\ c \end{pmatrix}_t = \begin{pmatrix} 0 & 1 & 0 \\ 0 & 0 & 1 \\ -(1+bc)/a^2 & c/a & b/a \end{pmatrix} \begin{pmatrix} a \\ b \\ c \end{pmatrix}_x, \qquad (5.5)$$

respectively.

We introduce the new variables

$$a = f_{xxx}, \quad b = f_{xxt}, \quad c = f_{xtt}, \quad d = f_{ttt}.$$

The compatibility conditions have the form

$$\begin{cases} a_t = b_x, \\ b_t = c_x, \\ c_t = d_x. \end{cases}$$

Moreover, equations (5.2) and (5.3) are equivalent to the relations

$$d = b^2 - ac$$

and

$$d = \frac{1 + bc}{a},$$

respectively. Thus, in the new variables equations (5.2) and (5.3), as the present author showed in [124], [125], become homogeneous 3×3 systems of hydrodynamic type:

$$\begin{cases} a_t = b_x, \\ b_t = c_x, \\ c_t = (b^2 - ac)_x, \end{cases} \qquad (5.6)$$

$$\begin{cases} a_t = b_x, \\ b_t = c_x, \\ c_t = ((1 + bc)/a)_x, \end{cases} \qquad (5.7)$$

respectively.

For second-order partial differential equations (which are of Monge–Ampère type) a similar transformation always results in a two-component system of hydrodynamic type. Any two-component system of hydrodynamic type can always be linearized, as is well known, by the classic hodograph method. In particular, in [155]

148   *O.I. MOKHOV*

the present author and Nutku considered the classic hyperbolic
Monge–Ampère equation

$$u_{tt}u_{xx} - u_{xt}^2 = -1,$$

which reduces to the integrable system of Chaplygin gas equations

$$\begin{cases} U_t = UU_x + V^{-3}V_x, \\ V_t = VU_x + UV_x \end{cases}$$

by the transformation

$$\begin{cases} U = u_{xt}/u_{xx}, \\ V = u_{xx}. \end{cases}$$

The main advantage of representing the equations of associativity in the form (5.6), (5.7) is the presence of an effective and
elaborate theory of integrability of one-dimensional homogeneous
systems of hydrodynamic type, which were considered here in Section 4, that is, systems of the form

$$u_t^i = v_j^i(u)u_x^j$$

(see, for example, the surveys of Tsarev [181], Dubrovin and Novikov
[36], and also the papers of Ferapontov [42]–[45], which are devoted
to non-diagonalizable systems of hydrodynamic type). Everywhere
in the present paper we consider only *strictly hyperbolic systems*,
that is, systems for which the eigenvalues of the corresponding matrices v_j^i are real and distinct. In particular, both the investigated
systems (5.6) and (5.7) are strictly hyperbolic.

The investigation made by the present author in [124], [125]
showed that both these systems (5.6) and (5.7) are non-diagonalizable (that is they do not possess Riemann invariants). This
simple fact was verified in different ways. In particular, it was
proved that the assumption on diagonalizability of the matrix of
the systems (5.6) and (5.7) leads to a contradiction after explicit
analytical calculations. Moreover, there is an effective criterion for
diagonalizability found by Haantjes [76].

Let $v_j^i(u)$ be an *affinor*, that is, a tensor of type $(1,1)$ on a
manifold M with local coordinates $(u^1, \ldots, u^N)$. By the *Nijenhuis*

tensor [156], [157] we mean the trivalent tensor $N_{ij}^k(u)$, which is
the *curvature tensor of the affinor* $v_j^i(u)$:

$$N_{ij}^k = v_i^s \frac{\partial v_j^k}{\partial u^s} - v_j^s \frac{\partial v_i^k}{\partial u^s} + v_s^k \frac{\partial v_i^s}{\partial u^j} - v_s^k \frac{\partial v_j^s}{\partial u^i}.$$

For an arbitrary vector field $\xi^i(u)$ on the manifold M, we have

$$N_{ij}^k \xi^i = \left(L_{v(\xi)}(v) - v L_\xi(v) \right)_j^k,$$

where L_ξ is the Lie derivative along the vector field $\xi(u)$.

A *Nijenhuis affinor* is an affinor such that its Nijenhuis tensor vanishes. In particular, any recursion operator defined by the quotient of two compatible Poisson structures (according to the Lenard–Magri scheme, see Section 2.1.6) is always a Nijenhuis affinor. The Nijenhuis tensor, which at present plays a very important role in the theory of integrable systems, in particular, in the theory of recursion operators, the theory of compatible Poisson structures, and the Lenard–Magri scheme (see [55], [58], [59], [67], [68], [80], [90]), was invented in an attempt to find a criterion for diagonalizability of the matrix $v_j^i(u)$ [156]. Later, the corresponding criterion was found by Haantjes [76].

We introduce the *Haantjes tensor*

$$H_{jk}^i = v_s^i v_r^s N_{jk}^r - v_s^i N_{rk}^s v_j^r - v_s^i N_{jr}^s v_k^r + N_{sr}^i v_j^s v_k^r.$$

If all the eigenvalues of the tensor $v_j^i(u)$ are real and distinct, then the tensor can be diagonalized in a neighbourhood of a point on the manifold if and only if its Haantjes tensor vanishes [76]. This criterion leads to huge calculations but all these calculations can be made by computer with the use of programs of symbol calculations. It has been checked that for the matrices of systems of hydrodynamic type (5.6) and (5.7) the Haantjes tensors are not equal to zero. Moreover, non-diagonalizability of the systems (5.6) and (5.7) follows from the fact that, as will be shown below, for the left eigenvectors $\xi_{(s)}(u)$:

$$\xi_{(s)i} v_j^i = \lambda_{(s)} \xi_{(s)j},$$

of the matrices $v_j^i(u)$ of these systems the Frobenius condition for the existence of an integrating multiplier

$$d(\xi_{(s)i} du^i) \wedge (\xi_{(s)k} du^k) = 0$$

is not fulfilled.

As Dubrovin noticed [30], the equation of associativity (5.1) is connected with the following spectral linear problem:

$$\frac{\partial \phi_\alpha}{\partial t^\beta} = z c^\gamma_{\alpha\beta} \phi_\gamma, \qquad z = \text{const.}$$

In fact, for this linear system the compatibility conditions have the form

$$z\frac{\partial c^\gamma_{\alpha\beta}}{\partial t^\lambda}\phi_\gamma + z^2 c^\gamma_{\alpha\beta} c^\varepsilon_{\gamma\lambda}\phi_\varepsilon = z\frac{\partial c^\gamma_{\alpha\lambda}}{\partial t^\beta}\phi_\gamma + z^2 c^\gamma_{\alpha\lambda} c^\varepsilon_{\gamma\beta}\phi_\varepsilon,$$

that is, they are equivalent to the equations of associativity (5.1) for the structural constants $c^\gamma_{\alpha\lambda}$ defined by the function $F(t^1,\ldots,t^N)$.

We consider the system (5.6) in more detail (we follow here to the paper of the present author and Ferapontov [153]). In our "hydrodynamical" variables a, b, c the spectral problem can be rewritten in the following form:

$$\Psi_x = zA\Psi = z\begin{pmatrix} 0 & 1 & 0 \\ b & a & 1 \\ c & b & 0 \end{pmatrix}\Psi,$$

$$\Psi_t = zB\Psi = z\begin{pmatrix} 0 & 0 & 1 \\ c & b & 0 \\ b^2-ac & c & 0 \end{pmatrix}\Psi. \qquad (5.8)$$

The compatibility conditions for the spectral problem (5.8) are equivalent to the following two relations between the matrices A and B:

$$\begin{cases} A_t = B_x, \\ [A,B] = 0, \end{cases} \qquad (5.9)$$

which are identically satisfied by virtue of equations (5.6) (here $[\cdot,\cdot]$ means the usual commutator of matrices).

Lemma 5.1 ([153]) *The eigenvalues of the matrix A are densities of conservation laws of the system (5.6).*

Thus, besides three evident conservation laws with the densities a, b, c, the system (5.6) also has three conservation laws with the densities u^1, u^2, u^3, which are roots of the characteristic equation

$$\det(\lambda E - A) = \lambda^3 - a\lambda^2 - 2b\lambda - c = 0.$$

By virtue of the obvious relation $a = u^1 + u^2 + u^3$, there are only five conservation laws among them with the densities u^1, u^2, u^3, b, c that are linear independent. The system (5.6) has no other conservation laws of hydrodynamic type, that is, with densities of the form $h(a, b, c)$.

In equations (5.6) we change from the "hydrodynamical" variables a, b, c to new dependent variables u^1, u^2, u^3, which are connected with a, b, c by the Vieta formulae

$$
\begin{cases}
a = u^1 + u^2 + u^3, \\
b = -\frac{1}{2}(u^1 u^2 + u^2 u^3 + u^3 u^1), \\
c = u^1 u^2 u^3.
\end{cases}
$$

To shorten the necessary calculations we note that the matrices A and B are connected by the relation

$$
B = A^2 - aA - bE.
$$

Consequently, the same relation is valid for the corresponding diagonal matrices $U = \mathrm{diag}(u^1, u^2, u^3)$ and $V = \mathrm{diag}(v^1, v^2, v^3)$:

$$
V = U^2 - aU - bE.
$$

Substituting the expressions for a and b and using the first equation of system (5.9), we obtain:

$$
U_t = (U^2 - (u^1 + u^2 + u^3)U + \frac{1}{2}(u^1 u^2 + u^2 u^3 + u^3 u^1)E)_x
$$

or, in components,

$$
\begin{pmatrix} u^1 \\ u^2 \\ u^3 \end{pmatrix}_t = \frac{1}{2} \begin{pmatrix} u^2 u^3 - u^1 u^2 - u^1 u^3 \\ u^1 u^3 - u^2 u^1 - u^2 u^3 \\ u^1 u^2 - u^3 u^1 - u^3 u^2 \end{pmatrix}_x =
$$

$$
= \frac{1}{2} \begin{pmatrix} 1 & -1 & -1 \\ -1 & 1 & -1 \\ -1 & -1 & 1 \end{pmatrix} \frac{d}{dx} \begin{pmatrix} \partial h/\partial u^1 \\ \partial h/\partial u^2 \\ \partial h/\partial u^3 \end{pmatrix}, \qquad (5.10)
$$

where $h = c = u^1 u^2 u^3$.

Hence, the system under consideration is Hamiltonian with the Hamiltonian operator

$$
M = \frac{1}{2} \begin{pmatrix} 1 & -1 & -1 \\ -1 & 1 & -1 \\ -1 & -1 & 1 \end{pmatrix} \frac{d}{dx} \qquad (5.11)
$$

and the Hamiltonian

$$H = \int c\,dx = \int u^1 u^2 u^3\,dx.$$

The density of momentum and the annihilators of the corresponding Poisson bracket have the following form:

$$2b = -u^1 u^2 - u^2 u^3 - u^3 u^1 \quad \text{(the density of momentum)},$$

$$u^1, u^2, u^3 \quad \text{(the annihilators)}.$$

In the initial variables a, b, c the Hamiltonian operator (5.11) has the form

$$M = \begin{pmatrix} -\frac{3}{2} & \frac{1}{2}a & b \\ \frac{1}{2}a & b & \frac{3}{2}c \\ b & \frac{3}{2}c & 2(b^2 - ac) \end{pmatrix} \frac{d}{dx} +$$

$$+ \begin{pmatrix} 0 & \frac{1}{2}a_x & b_x \\ 0 & \frac{1}{2}b_x & c_x \\ 0 & \frac{1}{2}c_x & (b^2 - ac)_x \end{pmatrix}, \tag{5.12}$$

that is, it defines a non-degenerate homogeneous Poisson structure of hydrodynamic type with a flat metric (a Dubrovin–Novikov bracket). In the papers of Tsarev [178], [181] an effective theory of integrability of diagonalizable Hamiltonian systems of hydrodynamic type, that is, Hamiltonian systems which can be reduced to Riemann invariants:

$$R_t^i = v^i(R)R_x^i,$$

was constructed. All these systems have infinitely many conservation laws and commuting flows of hydrodynamic type and can be integrated by the generalized hodograph method. However, as was shown above, the system (5.6) does not possess Riemann invariants. This explains, in particular, the fact that the system (5.6) has only finitely many conservation laws of hydrodynamic type. The general theory of integrability of non-diagonalizable Hamiltonian systems of hydrodynamic type (that is, systems which do not possess Riemann invariants) was developed in the papers of Ferapontov [42]–[45]. For three-component systems final results were obtained.

Theorem 5.2 ([43], [44]) *A non-diagonalizable Hamiltonian (with non-degenerate Poisson bracket) 3×3 system of hydrodynamic type is integrable if and only if it is linearly degenerate.*

We recall that a system of hydrodynamic type

$$u_t^i = v_j^i(u)u_x^j, \qquad i,j = 1,\dots,n, \qquad (5.13)$$

is called *linearly degenerate* if for the eigenvalues $\lambda^i(u)$ of the matrix $v_j^i(u)$ for any $i = 1,\dots,n$ the relations $L_{\vec{X}^i}(\lambda^i) = 0$, where $L_{\vec{X}^i}$ is the Lie derivative along the eigenvector $\vec{X}^i$ corresponding to the eigenvalue λ^i, are satisfied. There exists a simple and effective criterion for linear degeneracy that does not appeal to eigenvalues and eigenvectors [43].

Proposition 5.1 ([43]) *A system of hydrodynamic type (5.13) is linearly degenerate if and only if*

$$(\mathrm{grad}\, f_1)v^{n-1} + (\mathrm{grad}\, f_2)v^{n-2} + \cdots + (\mathrm{grad}\, f_n)E = 0,$$

where f_i are the coefficients of the characteristic polynomial

$$\det(\lambda \delta_j^i - v_j^i(u)) = \lambda^n + f_1(u)\lambda^{n-1} + f_2(u)\lambda^{n-2} + \cdots + f_n(u),$$

and v^n denotes the n-th power of the matrix v_j^i $(E = v^0)$.

It is easy to check that the system (5.6) is linearly degenerate. For the system (5.10) under consideration the eigenvalues λ^i and the left eigencovectors $\vec{l}^i$ corresponding to them have the form:

$$\lambda^1 = -u^1, \qquad \vec{l}^1 = (u^2 - u^3, u^1 - u^3, u^2 - u^1),$$

$$\lambda^2 = -u^2, \qquad \vec{l}^2 = (u^2 - u^3, u^1 - u^3, u^1 - u^2),$$

$$\lambda^3 = -u^3, \qquad \vec{l}^3 = (u^2 - u^3, u^3 - u^1, u^2 - u^1),$$

whence by virtue of the Frobenius criterion it follows that this system is non-diagonalizable (the Frobenius condition for the existence of an integrating multiplier is not satisfied).

Let $B(u) dx + A(u) dt$ and $N(u) dx + M(u) dt$ be two conservation laws of hydrodynamic type for a system of hydrodynamic type (5.13), that is, the differential forms are closed along solutions of this system. We change from the variables x, t to new independent variables $\tilde{x}, \tilde{t}$ with the help of the following relations (these transformations are called *reciprocal transformations*)

$$d\tilde{x} = B\,dx + A\,dt, \qquad d\tilde{t} = N\,dx + M\,dt. \qquad (5.14)$$

Then the system (5.13) is transformed to the form

$$u_{\tilde{t}}^i = \tilde{v}_j^i(u)u_{\tilde{x}}^j,$$

where the matrix $\tilde{v}$ is connected with the matrix v by the relation

$$\tilde{v} = (Bv - AE)(ME - Nv)^{-1}.$$

Theorem 5.3 ([43], [44]) *If a 3×3 system of hydrodynamic type (5.13) is linearly degenerate and Hamiltonian (with non-degenerate Poisson bracket of hydrodynamic type), then there exists a pair of conservation laws (5.14) of this system such that the corresponding transformed system has constant eigenvalues $\tilde{\lambda}^i$, moreover, without loss of generality we can assume that $\tilde{\lambda}^1 = 1$, $\tilde{\lambda}^2 = -1$, $\tilde{\lambda}^3 = 0$.*

For the system (5.10) the transformation (5.14), the existence of which is established by this theorem, has the following form:

$$d\tilde{x} = B\,dx + A\,dt = (u^1 - u^2)\,dx + u^3(u^2 - u^1)\,dt,$$
$$d\tilde{t} = N\,dx + M\,dt = (2u^3 - u^1 - u^2)\,dx +$$
$$+(2u^1u^2 - u^1u^3 - u^2u^3)\,dt. \qquad (5.15)$$

The transformed eigenvalues are equal to $1, -1, 0$, respectively. In the papers of Ferapontov [43], [44] it is proved that any non-diagonalizable, linearly degenerate, and Hamiltonian (with non-degenerate local Poisson structure of hydrodynamic type) homogeneous 3×3 system of hydrodynamic type reduces to the 3-wave system, which is integrated by the inverse problem method, by a reciprocal transformation of type (5.14) and a differential substitution of the first order. Applying this general construction of Ferapontov to our system (5.10), we obtain the following explicit relations connecting equations (5.10) and the 3-wave system.

1. The change from x, t to the new independent variables $\tilde{x}, \tilde{t}$ according to (5.15).

2. The change of the field variables from u^1, u^2, u^3 to p^1, p^2, p^3 according to the following relations:

$$p^1 = \frac{(u^2 - u^3)u^1_{\tilde{x}} + (u^1 - u^3)u^2_{\tilde{x}} + (u^2 - u^1)u^3_{\tilde{x}}}{2(u^2 - u^3)\sqrt{(u^2 - u^1)(u^3 - u^1)}},$$

$$p^2 = \frac{(u^2 - u^3)u^1_{\tilde{x}} + (u^1 - u^3)u^2_{\tilde{x}} + (u^1 - u^2)u^3_{\tilde{x}}}{2(u^3 - u^1)\sqrt{(u^2 - u^1)(u^2 - u^3)}}, \quad (5.16)$$

$$p^3 = \frac{(u^2 - u^3)u^1_{\tilde{x}} + (u^3 - u^1)u^2_{\tilde{x}} + (u^2 - u^1)u^3_{\tilde{x}}}{2(u^2 - u^1)\sqrt{(u^3 - u^1)(u^2 - u^3)}}.$$

Then we obtain the integrable 3-wave system

$$\begin{cases} p^1_t - p^1_{\tilde{x}} = -p^2 p^3, \\ p^2_t + p^2_{\tilde{x}} = -p^1 p^3, \\ p^3_t = -2p^1 p^2. \end{cases} \quad (5.17)$$

Any solution of the integrable 3-wave system (5.17) generates a three-parameter family of solutions of the equations of associativity (5.2).

Let us turn to another system of equations of associativity, namely, to the system (5.7). The spectral problem corresponding to the system (5.7) has the form

$$\Psi_x = zA\Psi = z \begin{pmatrix} 0 & 1 & 0 \\ 0 & b & a \\ 1 & c & b \end{pmatrix} \Psi,$$

$$\Psi_t = zB\Psi = z \begin{pmatrix} 0 & 0 & 1 \\ 1 & c & b \\ 0 & (1+bc)/a & c \end{pmatrix} \Psi. \quad (5.18)$$

It is easy to verify that the matrix B is related to A by

$$B = \frac{1}{a}(A^2 - bA).$$

The compatibility condition of the spectral problem (5.18)

$$A_t = B_x,$$

rewritten in terms of the eigenvalues of the matrices A and B (see Lemma 5.1), has the form

$$w_t^i = \left(\frac{1}{a} ((w^i)^2 - bw^i) \right)_x , \qquad (5.19)$$

where w^i are the eigenvalues of the matrix A, that is, the roots of the characteristic equation

$$\det(\lambda E - A) = \lambda^3 - 2b\lambda^2 + (b^2 - ac)\lambda - a = 0.$$

Expressing a and b by the Vieta formulae

$$\begin{cases} b = \frac{1}{2}(w^1 + w^2 + w^3), \\ a = w^1 w^2 w^3 \end{cases}$$

and substituting these expressions in (5.19), we obtain the explicit representation of equations (5.7) in coordinates w^i:

$$\begin{pmatrix} w^1 \\ w^2 \\ w^3 \end{pmatrix}_t = \frac{1}{2} \begin{pmatrix} (w^1 - w^2 - w^3)/w^2 w^3 \\ (w^2 - w^1 - w^3)/w^1 w^3 \\ (w^3 - w^1 - w^2)/w^1 w^2 \end{pmatrix}_x . \qquad (5.20)$$

We note that the integrable systems of hydrodynamic type (5.7) and (5.20) do not possess local Hamiltonian structures of hydrodynamic type (Poisson brackets of Dubrovin–Novikov type). Hamiltonian structures of hydrodynamic type corresponding to them are non-local (see Section 4.2).

We exhibit now the explicit relation between the systems (5.10) and (5.20). For this reason in equations (5.10) we change from x, t to the new independent variables $\tilde{x}$, $\tilde{t}$ according to the following relations:

$$d\tilde{x} = -\frac{1}{2}(u^1 u^2 + u^1 u^3 + u^2 u^3)\, dx + u^1 u^2 u^3\, dt, \quad d\tilde{t} = dx. \quad (5.21)$$

After the transformation (5.21) the system (5.10) has the form

$$\begin{pmatrix} 1/u^1 \\ 1/u^2 \\ 1/u^3 \end{pmatrix}_{\tilde{t}} = \frac{1}{2} \begin{pmatrix} (u^2 u^3)/u^1 - u^2 - u^3 \\ (u^1 u^3)/u^2 - u^1 - u^3 \\ (u^1 u^2)/u^3 - u^1 - u^2 \end{pmatrix}_{\tilde{x}} ,$$

which, as is easy to see, coincides with (5.20) after the transformation

$$w^i = \frac{1}{u^i}. \qquad (5.22)$$

In terms of the initial equations (5.2) and (5.3) the transformations (5.21) and (5.22) can be represented in the following way: the equation

$$f_{ttt} = f_{xxt}^2 - f_{xxx}f_{xtt}$$

is transformed into the equation

$$\tilde{f}_{\tilde{x}\tilde{x}\tilde{x}}\tilde{f}_{\tilde{t}\tilde{t}\tilde{t}} - \tilde{f}_{\tilde{x}\tilde{x}\tilde{t}}\tilde{f}_{\tilde{x}\tilde{t}\tilde{t}} = 1$$

under the transformation

$$\tilde{x} = f_{xt}, \quad \tilde{t} = x, \quad \tilde{f}_{\tilde{x}\tilde{x}} = t, \quad \tilde{f}_{\tilde{x}\tilde{t}} = -f_{xx}, \quad \tilde{f}_{\tilde{t}\tilde{t}} = f_{tt}. \qquad (5.23)$$

This transformation connecting solutions of the equations of associativity (5.2) and (5.3) is similar in form to the autotransformations of the associativity equations which were found by Dubrovin [29] (transformations of Legendre type) but, in contrast to them, they do not preserve the metric $\eta_{\alpha\beta}$.

Note that, considering solutions of the system (5.10), which are linear with respect to the variable x:

$$u^i = a^i(t)x,$$

we obtain a reduction of the system (5.10) to the well-known finite-dimensional dynamical Halphen system

$$\begin{cases} \dot{a}^1 = a^2 a^3 - a^1(a^2 + a^3), \\ \dot{a}^2 = a^1 a^3 - a^2(a^1 + a^3), \\ \dot{a}^3 = a^1 a^2 - a^3(a^1 + a^2). \end{cases} \qquad (5.24)$$

5.2 Poisson and symplectic structures of the equations of associativity

For the system of equations of associativity (5.4) the first Poisson structure generated by a flat metric was presented above in Section

5.1 [153]:

$$M_1 = \begin{pmatrix} -\frac{3}{2} & \frac{1}{2}a & b \\ \frac{1}{2}a & b & \frac{3}{2}c \\ b & \frac{3}{2}c & 2(b^2 - ac) \end{pmatrix} \frac{d}{dx} +$$

$$+ \begin{pmatrix} 0 & \frac{1}{2}a_x & b_x \\ 0 & \frac{1}{2}b_x & c_x \\ 0 & \frac{1}{2}c_x & (b^2 - ac)_x \end{pmatrix}, \qquad (5.25)$$

and the corresponding Hamiltonian of hydrodynamic type is

$$H = \int c \, dx.$$

The second Poisson structure of this system, which is compatible with the first one, is constructed from a Lagrangian found for equation (5.2) by Galvão and Nutku. It is a homogeneous differential-geometric Poisson structure of the third order (Dubrovin–Novikov type) with a flat metric, and this Poisson structure cannot be reduced to a constant form by a local change of coordinates on the manifold [50]:

$$M_2 = \begin{pmatrix} 0 & 0 & 1 \\ 0 & 1 & -a \\ 1 & -a & a^2 + 2b \end{pmatrix} \left(\frac{d}{dx}\right)^3 +$$

$$+ \begin{pmatrix} 0 & 0 & 0 \\ 0 & 0 & -2a_x \\ 0 & -a_x & 3(b_x + aa_x) \end{pmatrix} \left(\frac{d}{dx}\right)^2 +$$

$$+ \begin{pmatrix} 0 & 0 & 0 \\ 0 & 0 & 0 \\ 0 & 0 & b_{xx} + a_x^2 + aa_{xx} \end{pmatrix} \frac{d}{dx}, \qquad (5.26)$$

the Hamiltonian is non-local and has the form:

$$H_1 = -\int \left(\frac{1}{2}a\left(\left(\frac{d}{dx}\right)^{-1}b\right)^2 + \right.$$

$$\left. + \left(\left(\frac{d}{dx}\right)^{-1}b\right)\left(\left(\frac{d}{dx}\right)^{-1}c\right)\right) dx. \qquad (5.27)$$

Further, according to the general bi-Hamiltonian Lenard–Magri scheme we can construct a recursion operator, to find higher integrals and higher equations of bi-Hamiltonian integrable hierarchy of the equation of associativity (5.2) [50].

We introduce the new variables

$$p = f_x, \quad q = f_t, \quad r = f_{tt}.$$

Then the equation of associativity (5.2) is equivalent to the system of evolution equations

$$\begin{cases} p_t = q_x, \\ q_t = r, \\ r_t = q_{xx}^2 - p_{xx} r_x, \end{cases} \tag{5.28}$$

for which Galvão and Nutku found a Lagrangian representation defined by the local action

$$S = \int [2 p_x q_{xx} p_t + (2 p_x p_{xx} - q_x) q_t +$$

$$+ 2 p r_t - 2 q r_x + q_x^2 p_{xx}] \, dx \, dt. \tag{5.29}$$

For the initial variable f the action generating the equation of associativity (5.2) has the form

$$S = \int (f_{xt}^2 f_{xxx} + f_{xt} f_{tt}) \, dx \, dt.$$

The system (5.28) is equivalent to the Lagrangian system

$$\frac{\delta S}{\delta u^i(x)} = 0$$

(here $u^1 = p$, $u^2 = q$, $u^3 = r$). The action (5.29) is degenerate and belongs to the class of actions (2.104) considered in Section 2.3, and generating local symplectic structures of the corresponding Lagrangian systems. The symplectic operator corresponding to the action (5.29) can be calculated by the general explicit formula (2.106) and has the form:

$$(M_{ij}) = \begin{pmatrix} q_{xx} \frac{d}{dx} + \frac{d}{dx} \circ q_{xx} & -p_{xx} \frac{d}{dx} & -1 \\ -\frac{d}{dx} \circ p_{xx} & -\frac{d}{dx} & 0 \\ 1 & 0 & 0 \end{pmatrix}. \tag{5.30}$$

And correspondingly, the Hamiltonian of the symplectic representation (2.105) for the system (5.28) has the form

$$H = \int \left[q_x r + \frac{1}{2} q_x^2 p_{xx} \right] dx.$$

The operator J^{ij}, which is inverse to the operator M_{ij}, is Hamiltonian:

$$(J^{ij}) = \qquad\qquad\qquad\qquad\qquad\qquad\qquad (5.31)$$

$$= \begin{pmatrix} 0 & 0 & 1 \\ 0 & -\left(\frac{d}{dx}\right)^{-1} & -p_{xx} \\ -1 & p_{xx} & \frac{d}{dx} \circ q_{xx} + q_{xx}\frac{d}{dx} + p_{xx}\frac{d}{dx} \circ p_{xx} \end{pmatrix}.$$

The Poisson structure (5.31) is non-local, but it is easy to change to
variables in which this Hamiltonian representation of the equation
of associativity (5.2) is local, that is, the Poisson structure, the
Hamiltonian, and the momentum, are all local. In fact, in the new
variables v^i:

$$v^1 = p_x = f_{xx}, \qquad v^2 = q_x = f_{xt}, \qquad v^3 = r = f_{tt},$$

the equation of associativity (5.2) is equivalent to the evolution
system

$$\begin{cases} v_t^1 = v_x^2, \\ v_t^2 = v_x^3, \\ v_t^3 = (v_x^2)^2 - v_x^1 v_x^3, \end{cases} \qquad (5.32)$$

which has the following local Hamiltonian representation:

$$\begin{pmatrix} v^1 \\ v^2 \\ v^3 \end{pmatrix}_t = \qquad\qquad\qquad\qquad\qquad\qquad (5.33)$$

$$= \begin{pmatrix} 0 & 0 & \frac{d}{dx} \\ 0 & \frac{d}{dx} & -\frac{d}{dx} \circ v_x^1 \\ \frac{d}{dx} & -v_x^1 \frac{d}{dx} & v_x^2 \frac{d}{dx} + \frac{d}{dx} \circ v_x^2 + v_x^1 \frac{d}{dx} \circ v_x^1 \end{pmatrix} \begin{pmatrix} \delta H/\delta v^1(x) \\ \delta H/\delta v^2(x) \\ \delta H/\delta v^3(x) \end{pmatrix},$$

where the Hamiltonian H has the form

$$H = \int \left(v^2 v^3 + \frac{1}{2} v_x^1 (v^2)^2 \right) dx.$$

The momentum P of the Poisson structure of the system (5.33) is also local:

$$P = \int \left(v^1 v^3 - \frac{1}{2}(v^1)^2 v_x^2 + \frac{1}{2}(v^2)^2 \right) dx,$$

$$\begin{pmatrix} v_x^1 \\ v_x^2 \\ v_x^3 \end{pmatrix} =$$

$$= \begin{pmatrix} 0 & 0 & \frac{d}{dx} \\ 0 & \frac{d}{dx} & -\frac{d}{dx} \circ v_x^1 \\ \frac{d}{dx} & -v_x^1 \frac{d}{dx} & v_x^2 \frac{d}{dx} + \frac{d}{dx} \circ v_x^2 + v_x^1 \frac{d}{dx} \circ v_x^1 \end{pmatrix} \begin{pmatrix} \delta P/\delta v^1(x) \\ \delta P/\delta v^2(x) \\ \delta P/\delta v^3(x) \end{pmatrix}.$$

We change now to our "hydrodynamical" variables a, b, c:

$$a = v_x^1 = f_{xxx}, \qquad b = v_x^2 = f_{xxt}, \qquad c = v_x^3 = f_{xtt}.$$

In these variables the Hamiltonian and the momentum become non-local but, on the other hand, the Poisson structure (5.31) in these variables is a homogeneous matrix differential operator of the third order (5.26), that is, we obtain a non-trivial example of a system of equations which possesses a homogeneous differential-geometric Poisson structure of the third order. It is interesting that the "hydrodynamical" variables a, b, c are special coordinates for this Poisson structure, that is, in these coordinates the last connection vanishes (see Section 4.5). According to the Potemin–Doyle criterion, which is presented in Section 4.5, the Poisson structure (5.26) cannot be reduced to constant form (4.155) by local changes of coordinates, since the last connection is not compatible with the metric g^{ij} of this Poisson structure. The metric g^{ij} is quadratic in the special coordinates:

$$(g^{ij}) = \begin{pmatrix} 0 & 0 & 1 \\ 0 & 1 & -a \\ 1 & -a & a^2 + 2b \end{pmatrix}. \tag{5.34}$$

It is interesting that the metric g^{ij} is flat. By a local change of variables

$$a = u, \qquad b = v - u^2, \qquad c = w + uv - \frac{u^3}{3}$$

the metric with lower indices

$$(g_{ij}) = \begin{pmatrix} -2b & a & 1 \\ a & 1 & 0 \\ 1 & 0 & 0 \end{pmatrix}, \qquad (5.35)$$

which is linear in the "hydrodynamical" variables a, b, c, reduces to the constant form

$$(\tilde{g}_{ij}) = \begin{pmatrix} 0 & 0 & 1 \\ 0 & 1 & 0 \\ 1 & 0 & 0 \end{pmatrix}. \qquad (5.36)$$

The Poisson structure (5.26) possesses non-local momentum

$$P = \int \left[-\left(\left(\frac{d}{dx} \right)^{-1} a \right) \left(\left(\frac{d}{dx} \right)^{-1} c \right) - \right.$$
$$\left. -\frac{1}{2} \left(\left(\frac{d}{dx} \right)^{-1} b \right)^2 + \frac{1}{2} b \left(\left(\frac{d}{dx} \right)^{-1} a \right)^2 \right] dx \qquad (5.37)$$

and non-local *Casimir functionals*, that is, functionals C_k whose variational derivatives belong to kernel of the Poisson structure:

$$M_2^{ij} \frac{\delta C_k}{\delta u^j(x)} \equiv 0,$$

$$C_1 = \int \left(\frac{d}{dx} \right)^{-1} a \, dx, \quad C_2 = \int \left(\frac{d}{dx} \right)^{-1} b \, dx,$$

$$C_3 = \int \left[\left(\frac{d}{dx} \right)^{-1} c + a \left(\frac{d}{dx} \right)^{-1} b \right] dx \qquad (5.38)$$

(the variables a, b, c are also the annihilators, that is, densities of Casimir functionals for the Poisson structure (5.26)).

As is shown above, the Poisson structure (5.26) cannot be reduced to a constant form by local changes of coordinates. In this case, in order to reduce the Poisson structure (5.26) to a constant form, it is necessary to choose as new variables w^i the densities of the non-local Casimir functionals C_k. This change is realized by a differential substitution of the second order:

$$a = w_x^1, \quad b = w_x^2, \quad c = w_x^3 - w_x^1 w_x^2 - w^2 w_{xx}^1.$$

In the Casimir variables w^i the Poisson structure (5.26) becomes constant:

$$(\widetilde{M}_2^{ij}) = - \begin{pmatrix} 0 & 0 & 1 \\ 0 & 1 & 0 \\ 1 & 0 & 0 \end{pmatrix} \frac{d}{dx}, \tag{5.39}$$

and the equation of associativity (5.2) becomes a canonical Hamiltonian system of diffusion type

$$\begin{cases} w_t^1 = w_x^2, \\ w_t^2 = w_x^3 - (w^2 w_x^1)_x, \\ w_t^3 = (w^2 w_x^2)_x \end{cases} \tag{5.40}$$

with the Poisson structure (5.39) and the local Hamiltonian

$$H = \int \left[\frac{1}{2}(w^2)^2 w_x^1 - w^2 w^3 \right] dx.$$

Theorem 5.4 ([50]) *The Poisson structures (5.25) and (5.26) are compatible and define the bi-Hamiltonian structure of the equations of associativity (5.2).*

The proof consists in checking by direct calculations the Jacobi identity for the sum of the local Poisson structures (5.25) and (5.26) (these calculations are also verified with the use of the program of symbol calculations REDUCE).

Thus, for the equation of associativity (5.2), the recursion operator $R = M_2(M_1)^{-1}$ generating a hierarchy of the equation (5.2) and raising the orders of higher commuting flows from the hierarchy and higher conservation laws is constructed. In the next section we consider the first higher conservation laws of equation (5.2) [50]. It turns out that the conservation laws are non-degenerate and we can apply the theorem on canonical Hamiltonian finite-dimensional reduction to a stationary manifold of non-degenerate integrals of a system [100], [104], [112] to the equation of associativity (5.2).

We present here one more example, which was found by the present author and Nutku, of a system arising in applications and possessing a homogeneous Poisson structure of the third order. This is the system of the Chaplygin gas equations

$$\begin{cases} U_t = UU_x + V^{-3}V_x, \\ V_t = VU_x + UV_x, \end{cases} \tag{5.41}$$

which we have already considered above and which is connected by the transformation

$$U = \frac{u_{xt}}{u_{xx}}, \quad V = u_{xx}$$

with the hyperbolic Monge–Ampère equation:

$$u_{tt}u_{xx} - u_{xt}^2 = -1.$$

In Section 4 we considered three local and one non-local compatible Poisson structures of hydrodynamic type for the system of the Chaplygin gas equations. Here we exhibit a homogeneous Poisson structure of the third order for this system [154]. In Section 2.3.3 we presented local symplectic and Poisson structures for the Monge–Ampère equation. The Monge–Ampère equation rewritten in the form of the system

$$\begin{cases} u_t = q, \\ q_t = \left(q_x^2 - 1\right)/u_{xx} \end{cases}$$

possesses the Poisson structure

$$K = \begin{pmatrix} 0 & \frac{1}{u_{xx}} \\ -\frac{1}{u_{xx}} & \frac{q_x}{u_{xx}^2}\frac{d}{dx} + \frac{d}{dx} \circ \frac{q_x}{u_{xx}^2} \end{pmatrix}$$

and the Hamiltonian

$$H = \int \left(\frac{1}{2}q^2 u_{xx} + u\right) dx.$$

We introduce more convenient "hydrodynamical" variables

$$a = u_{xx} = V, \quad b = q_x = u_{xt} = UV,$$

in which the Chaplygin gas equations have the form:

$$a_t = b_x,$$

$$b_t = -\frac{b^2 - 1}{a^2}a_x + 2\frac{b}{a}b_x.$$

The Poisson structure of the Monge–Ampère equation is transformed into the homogeneous Poisson structure of the third order for the Chaplygin gas:

$$J_1 = \begin{pmatrix} \frac{d^2}{dx^2} & 0 \\ 0 & \frac{d}{dx} \end{pmatrix} \begin{pmatrix} 0 & \frac{1}{a} \\ -\frac{1}{a} & \frac{b}{a^2}\frac{d}{dx} + \frac{d}{dx} \circ \frac{b}{a^2} \end{pmatrix} \begin{pmatrix} \frac{d^2}{dx^2} & 0 \\ 0 & -\frac{d}{dx} \end{pmatrix}.$$

The variables a, b are special for this homogeneous Poisson structure (the last connection vanishes):

$$
J_1 = \begin{pmatrix} 0 & -\frac{1}{a} \\ -\frac{1}{a} & -\frac{2b}{a^2} \end{pmatrix} \frac{d^3}{dx^3} + \begin{pmatrix} 0 & 2\frac{a_x}{a^2} \\ \frac{a_x}{a^2} & 6\frac{ba_x}{a^3} - 3\frac{b_x}{a^2} \end{pmatrix} \frac{d^2}{dx^2} +
$$

$$
+ \begin{pmatrix} 0 & \frac{a_{xx}}{a^2} - 2\frac{a_x^2}{a^3} \\ 0 & 2\frac{ba_{xx}}{a^3} + 4\frac{b_x a_x}{a^3} - 6\frac{ba_x^2}{a^4} - \frac{b_{xx}}{a^2} \end{pmatrix} \frac{d}{dx}.
$$

The metric

$$
(g^{ij}) = \begin{pmatrix} 0 & -1/a \\ -1/a & -2b/a^2 \end{pmatrix}
$$

is a flat metric but according to the Potemin–Doyle criterion this Poisson structure cannot be reduced to a constant form by a local change of coordinates on the manifold (otherwise it would be a constant structure in the special coordinates). We show that the metric is flat:

$$
(g_{ij}) = \begin{pmatrix} 2b & -a \\ -a & 0 \end{pmatrix},
$$

$$
ds^2 = 2b\,da^2 - 2a\,da\,db = 2\,da(b\,da - a\,db) =
$$

$$
= 2a^2\,da(-d(b/a)) = -\frac{2}{3}d(a^3)d(b/a).
$$

Thus, by the local change of variables

$$
u^1 = \frac{1}{3}a^3, \qquad u^2 = -\frac{b}{a}
$$

the metric g_{ij} reduces to a constant form

$$
(\tilde{g}_{ij}) = \begin{pmatrix} 0 & 1 \\ 1 & 0 \end{pmatrix}.
$$

The Poisson structure reduces to a constant form with the help of a differential substitution of the second order. In fact, the non-local functionals

$$
C_1 = \int \left(\frac{d}{dx} \right)^{-1} a\,dx, \qquad C_2 = \int a \left(\frac{d}{dx} \right)^{-1} b\,dx,
$$

$$
C_1 = \int \left(\left(\frac{d}{dx} \right)^{-1} a \right)^2 dx \tag{5.42}
$$

are Casimir functionals for the Poisson structure. Choosing as new variables the densities of two first Casimir functionals, we obtain

$$v^1 = \left(\frac{d}{dx}\right)^{-1} a, \qquad v^2 = a\left(\frac{d}{dx}\right)^{-1} b,$$

or

$$a = v_x^1, \qquad b = \frac{v_x^2}{v_x^1} - \frac{v^2 v_{xx}^1}{(v_x^1)^2}.$$

In the variables (v^1, v^2) the Poisson structure becomes constant:

$$\tilde{J}_1 = \left(\begin{array}{cc} 0 & 1 \\ 1 & 0 \end{array}\right)\frac{d}{dx}.$$

Let us return to the equations of associativity. In the general case, for an arbitrary N the equations of associativity reduce to $N - 2$ commuting evolution one-dimensional homogeneous non-diagonalizable $N(N - 1)/2$-component integrable systems of hydrodynamic type.

We consider in detail the case $N = 4$ for the antidiagonal metric [51]:

$$\eta_{\alpha\beta} = \left(\begin{array}{cccc} 0 & 0 & 0 & 1 \\ 0 & 0 & 1 & 0 \\ 0 & 1 & 0 & 0 \\ 1 & 0 & 0 & 0 \end{array}\right). \tag{5.43}$$

In this case the function F has the form:

$$F = \frac{1}{2}(t^1)^2 t^4 + t^1 t^2 t^3 + f(t^2, t^3, t^4)$$

and the corresponding equations of associativity reduce to the following complicated overdetermined system:

$$-2f_{xyz} - f_{xyy}f_{xxy} + f_{yyy}f_{xxx} = 0,$$
$$-f_{xzz} - f_{xyy}f_{xxz} + f_{yyz}f_{xxx} = 0,$$
$$-2f_{xyz}f_{xxz} + f_{xzz}f_{xxy} + f_{yzz}f_{xxx} = 0, \tag{5.44}$$
$$f_{zzz} - (f_{xyz})^2 + f_{xzz}f_{xyy} - f_{yyz}f_{xxz} + f_{yzz}f_{xxy} = 0,$$
$$f_{yyy}f_{xzz} - 2f_{yyz}f_{xyz} + f_{yzz}f_{xyy} = 0,$$

where $x = t^2$, $y = t^3$, $z = t^4$.

After introducing the new variables

$$f_{xxx} = a, \qquad f_{xxy} = b, \qquad f_{xxz} = c,$$

$$f_{xyy} = d, \qquad f_{xyz} = e, \qquad f_{xzz} = f$$

the system (5.44) can be rewritten as a pair of commuting 6×6 systems of hydrodynamic type:

$$
\begin{pmatrix} a \\ b \\ c \\ d \\ e \\ f \end{pmatrix}_y = \begin{pmatrix} b \\ d \\ e \\ R \\ P \\ S \end{pmatrix}_x, \qquad (5.45)
$$

$$
\begin{pmatrix} a \\ b \\ c \\ d \\ e \\ f \end{pmatrix}_z = \begin{pmatrix} c \\ e \\ f \\ P \\ S \\ Q \end{pmatrix}_x, \qquad (5.46)
$$

where we introduce the following notation:

$$f_{yyz} = P = \frac{cd + f}{a}, \quad f_{yyy} = R = \frac{2e + bd}{a},$$

$$f_{yzz} = S = \frac{2ec - bf}{a},$$

$$f_{zzz} = Q = e^2 - fd + \frac{c^2 d + cf - 2bec + b^2 f}{a}.$$

The systems of equations (5.45) and (5.46) are connected with the following spectral problem:

$$\Psi_x = \lambda A \Psi, \qquad \Psi_y = \lambda B \Psi, \qquad \Psi_z = \lambda C \Psi, \qquad (5.47)$$

where the matrices A, B, C are defined by the relations

$$
A = \begin{pmatrix} 0 & 1 & 0 & 0 \\ c & b & a & 0 \\ e & d & b & 1 \\ f & e & c & 0 \end{pmatrix}, \qquad
B = \begin{pmatrix} 0 & 0 & 1 & 0 \\ e & d & b & 1 \\ P & R & d & 0 \\ S & P & e & 0 \end{pmatrix},
$$

$$C = \begin{pmatrix} 0 & 0 & 0 & 1 \\ f & e & c & 0 \\ S & P & e & 0 \\ Q & S & f & 0 \end{pmatrix}.$$ (5.48)

The compatibility conditions for the spectral problem (5.47) lead to the condition of commutativity of the matrices A, B, C:

$$[A, B] = [A, C] = [B, C] = 0,$$

and to the equations

$$A_y = B_x, \quad A_z = C_x, \quad B_z = C_x,$$

which are identically satisfied by virtue of equations (5.45) and (5.46).

We consider the eigenvalues u^i of the matrix A, which are the roots of the characteristic polynomial of the fourth degree

$$\det(A - \lambda E) = \lambda^4 - 2b\lambda^3 + (b^2 - ad - 2c)\lambda^2 +$$
$$+2(bc - ae)\lambda + c^2 - af = 0.$$ (5.49)

The variables b, c, e, f are connected with u^1, u^2, u^3, u^4, a, d by the Vieta formulae:

$$b = \frac{1}{2}(u^1 + u^2 + u^3 + u^4),$$

$$c = \frac{1}{4}((u^1)^2 + (u^2)^2 + (u^3)^2 + (u^4)^2) -$$
$$-\frac{1}{8}(u^1 + u^2 + u^3 + u^4)^2 - \frac{1}{2}ad,$$

$$e = \frac{bc + u^1u^2u^3 + u^1u^2u^4 + u^1u^3u^4 + u^2u^3u^4}{a},$$

$$f = \frac{c^2 - u^1u^2u^3u^4}{a}.$$

In the new variables a, u^1, u^2, u^3, u^4, d both systems (5.45) and (5.46) can be represented in Hamiltonian forms

$$
\begin{pmatrix} a \\ u^1 \\ u^2 \\ u^3 \\ u^4 \\ d \end{pmatrix}_y = \tag{5.50}
$$

$$
= \begin{pmatrix} 0 & 0 & 0 & 0 & 0 & -2 \\ 0 & 1 & -1 & -1 & -1 & 0 \\ 0 & -1 & 1 & -1 & -1 & 0 \\ 0 & -1 & -1 & 1 & -1 & 0 \\ 0 & -1 & -1 & -1 & 1 & 0 \\ -2 & 0 & 0 & 0 & 0 & 0 \end{pmatrix} \frac{d}{dx} \begin{pmatrix} \partial e/\partial a \\ \partial e/\partial u^1 \\ \partial e/\partial u^2 \\ \partial e/\partial u^3 \\ \partial e/\partial u^4 \\ \partial e/\partial d \end{pmatrix},
$$

with the Hamiltonian

$$
H = \int e\, dx,
$$

and

$$
\begin{pmatrix} a \\ u^1 \\ u^2 \\ u^3 \\ u^4 \\ d \end{pmatrix}_z = \tag{5.51}
$$

$$
= \frac{1}{2} \begin{pmatrix} 0 & 0 & 0 & 0 & 0 & -2 \\ 0 & 1 & -1 & -1 & -1 & 0 \\ 0 & -1 & 1 & -1 & -1 & 0 \\ 0 & -1 & -1 & 1 & -1 & 0 \\ 0 & -1 & -1 & -1 & 1 & 0 \\ -2 & 0 & 0 & 0 & 0 & 0 \end{pmatrix} \frac{d}{dx} \begin{pmatrix} \partial f/\partial a \\ \partial f/\partial u^1 \\ \partial f/\partial u^2 \\ \partial f/\partial u^3 \\ \partial f/\partial u^4 \\ \partial f/\partial d \end{pmatrix},
$$

with the Hamiltonian

$$
H = \int \frac{1}{2} f\, dx,
$$

respectively.

The characteristic velocities of the systems (5.45) and (5.46) are defined by the relations

$$\lambda^1 = (u^1 + u^2 - u^3 - u^4)/(2a), \qquad \lambda^2 = -\lambda^1,$$

$$\lambda^3 = (u^1 - u^2 + u^3 - u^4)/(2a), \qquad \lambda^4 = -\lambda^3,$$

$$\lambda^5 = (-u^1 + u^2 + u^3 - u^4)/(2a), \qquad \lambda^6 = -\lambda^5$$

and

$$\mu^1 = \mu + (u^3 u^4)/a, \qquad \mu^2 = \mu + (u^1 u^2)/a,$$

$$\mu^3 = \mu + (u^2 u^4)/a, \qquad \mu^4 = \mu + (u^1 u^3)/a,$$

$$\mu^5 = \mu + (u^1 u^4)/a, \qquad \mu^6 = \mu + (u^2 u^3)/a,$$

where

$$\mu = -\frac{1}{2}d + \frac{(u^1)^2 + (u^2)^2 + (u^3)^2 + (u^4)^2}{8a}$$
$$- \frac{u^1 u^2 + u^1 u^3 + u^1 u^4 + u^2 u^3 + u^2 u^4 + u^3 u^4}{4a},$$

respectively.

We present here also the form of the constant non-degenerate Poisson structure

$$M = \begin{pmatrix} 0 & 0 & 0 & 0 & 0 & -2 \\ 0 & 1 & -1 & -1 & -1 & 0 \\ 0 & -1 & 1 & -1 & -1 & 0 \\ 0 & -1 & -1 & 1 & -1 & 0 \\ 0 & -1 & -1 & -1 & 1 & 0 \\ -2 & 0 & 0 & 0 & 0 & 0 \end{pmatrix} \frac{d}{dx}$$

in the initial variables a, b, c, d, e, f:

$$M = \begin{pmatrix} 0 & 0 & a & -2 & b & 2c \\ 0 & -2 & b & 0 & d & 2e \\ a & b & 2c & d & 2e & 3f \\ -2 & 0 & d & 0 & R & 2P \\ b & d & 2e & R & 2P & 3S \\ 2c & 2e & 3f & 2P & 3S & 4Q \end{pmatrix} \frac{d}{dx} +$$

$$+ \begin{pmatrix} 0 & 0 & a_x & 0 & b_x & 2c_x \\ 0 & 0 & b_x & 0 & d_x & 2e_x \\ 0 & 0 & c_x & 0 & e_x & 2f_x \\ 0 & 0 & d_x & 0 & R_x & 2P_x \\ 0 & 0 & e_x & 0 & P_x & 2S_x \\ 0 & 0 & f_x & 0 & S_x & 2Q_x \end{pmatrix}. \tag{5.52}$$

Both the investigated systems (5.45) and (5.46) can be transformed to the integrable 6-wave system.

To conclude this section we consider a more general class of equations of associativity in which it is not assumed that the condition 1 defining the metric $\eta_{\alpha\beta}$ via the function $F(t^1, \ldots, t^N)$ is fulfilled, that is, now $\eta_{\alpha\beta}$ is an arbitrary constant non-degenerate symmetric matrix, with the help of which the indices of the structural constants are raised and lowered. This means that we do not assume the presence of units in the associative algebras $A(t)$. As before, the metric $\eta_{\alpha\beta}$ defines an invariant non-degenerate scalar product on the commutative associative algebras $A(t)$. These equations of associativity arise already for $N = 2$ and they were also considered by Witten in [189], [190]:

$$F_{t^1 t^1 t^1} F_{t^2 t^2 t^2} = F_{t^1 t^1 t^2} F_{t^1 t^2 t^2} \tag{5.53}$$

if

$$(\eta_{ij}) = \begin{pmatrix} 0 & 1 \\ 1 & 0 \end{pmatrix},$$

and

$$F_{t^1 t^1 t^1} F_{t^1 t^2 t^2} + F_{t^1 t^1 t^2} F_{t^2 t^2 t^2} = \left(F_{t^1 t^1 t^2}\right)^2 + \left(F_{t^1 t^2 t^2}\right)^2 \tag{5.54}$$

if

$$(\eta_{ij}) = \begin{pmatrix} 1 & 0 \\ 0 & 1 \end{pmatrix}.$$

Theorem 5.5 ([125], [126]) *The equations (5.53) and (5.54) are equivalent to integrable non-diagonalizable homogeneous systems of hydrodynamic type*

$$\begin{pmatrix} a \\ b \\ c \end{pmatrix}_t = \begin{pmatrix} 0 & 1 & 0 \\ 0 & 0 & 1 \\ -bc/a^2 & c/a & b/a \end{pmatrix} \begin{pmatrix} a \\ b \\ c \end{pmatrix}_x, \tag{5.55}$$

$$\begin{pmatrix} a \\ b \\ c \end{pmatrix}_t = \tag{5.56}$$

$$= \begin{pmatrix} 0 & 1 & 0 \\ 0 & 0 & 1 \\ -c/b & 1 - c^2/b^2 + ac/b^2 & 2c/b - a/b \end{pmatrix} \begin{pmatrix} a \\ b \\ c \end{pmatrix}_x ,$$

respectively.

We consider in more detail here the Witten equation (5.53). Introducing, as before, the new variables

$$a = F_{xxx}, \quad b = F_{xxt}, \quad c = F_{xtt},$$

where $x = t^1$, $t = t^2$, we can rewrite equation (5.53) as a homogeneous system of hydrodynamic type

$$\begin{cases} a_t = b_x, \\ b_t = c_x, \\ c_t = (bc/a)_x. \end{cases} \tag{5.57}$$

Although the system (5.57) is similar to the system (5.7) investigated above, this system (5.57) is actually considerably simpler: in the joint paper of Ferapontov and the present author [52] it is shown that this system is linearized by a chain of transformations. We introduce new variables

$$R^1 = \frac{\sqrt{c}}{\sqrt{a}}, \quad R^2 = b - \sqrt{ac}, \quad R^3 = b + \sqrt{ac}.$$

In the variables R^i the system (5.57) has the form

$$R_t^1 = R^1 \frac{R^3 + R^2}{R^3 - R^2} R_x^1,$$
$$R_t^2 = -(R^1 R^2)_x, \tag{5.58}$$
$$R_t^3 = (R^1 R^3)_x,$$

whence it follows that R^1 is a Riemann invariant (in contrast to the system (5.7), which has no Riemann invariants). Let us consider the reciprocal transformation

$$\begin{cases} d\tilde{x} = (R^3 - R^2)\, dx + R^1(R^3 + R^2)\, dt, \\ d\tilde{t} = (R^3 + R^2)\, dx + R^1(R^3 - R^2)\, dt. \end{cases} \tag{5.59}$$

In the variables $\tilde{x}$, $\tilde{t}$ the system (5.59) has the form

$$\begin{cases} R_{\tilde{t}}^1 = 0, \\ R_{\tilde{t}}^2 = -R_{\tilde{x}}^2 - \frac{R^2(R^3-R^2)}{2R^1R^3}R_{\tilde{x}}^1, \\ R_{\tilde{t}}^3 = R_{\tilde{x}}^3 - \frac{R^3(R^3-R^2)}{2R^1R^2}R_{\tilde{x}}^1. \end{cases} \quad (5.60)$$

After the transformation (5.59) the eigenvalues of the system (5.59), which are

$$\lambda^1 = \frac{R^1(R^3+R^2)}{R^3-R^2}, \quad \lambda^2 = R^1, \quad \lambda^3 = -R^1,$$

become constant: $\lambda^1 = 0$, $\lambda^2 = 1$, $\lambda^3 = -1$, respectively. In the joint paper of Ferapontov and the author [52] it is shown that the system (5.60) reduces to the following form:

$$\begin{cases} p_{\tilde{t}}^1 = 0, \\ p_{\tilde{t}}^2 = p_{\tilde{x}}^2 - \frac{1}{2}p^1(p^2+p^3), \\ p_{\tilde{t}}^3 = -p_{\tilde{x}}^3 + \frac{1}{2}p^1(p^2+p^3), \end{cases} \quad (5.61)$$

and the variables p^i and R^i are connected by the relations

$$p^1 = \frac{R_{\tilde{x}}^1}{R^1},$$

$$p^2 = \left(\frac{1}{R^3} - \frac{1}{R^2}\right)\frac{R_{\tilde{x}}^1}{R^1} + 2\frac{R_{\tilde{x}}^3}{(R^3)^2},$$

$$p^3 = \left(\frac{1}{R^2} - \frac{1}{R^3}\right)\frac{R_{\tilde{x}}^1}{R^1} + 2\frac{R_{\tilde{x}}^2}{(R^2)^2}, \quad (5.62)$$

that is, the change from R^i to p^i is a differential substitution of the first order.

The system of equations (5.61) is really a linear system with the general solution

$$p^1 = \phi(\tilde{x}), \qquad p^2 = \frac{1}{2}(g_{\tilde{x}} + g_{\tilde{t}}), \qquad p^3 = \frac{1}{2}(g_{\tilde{x}} - g_{\tilde{t}}), \quad (5.63)$$

where $\phi(\tilde{x})$ is an arbitrary function of the variable $\tilde{x}$ and $g(\tilde{x}, \tilde{t})$ is an arbitrary solution of the linear wave equation

$$g_{\tilde{t}\tilde{t}} - g_{\tilde{x}\tilde{x}} + \phi(\tilde{x})g_{\tilde{x}} = 0. \quad (5.64)$$

5.3 Theorem on a canonical Hamiltonian representation of the restriction of an arbitrary evolution system to the set of stationary points of its non-degenerate integral and its applications to the equations of associativity and systems of hydrodynamic type

In this section we present the following absolutely general theorem: the restriction of any one-dimensional evolution system to the stationary manifold of its non-degenerate integral is always a canonical Hamiltonian finite-dimensional dynamical system such that its Hamiltonian is always in involution with the Hamiltonian of the Lagrangian system for the stationary points of the integral under consideration [100], [104]. This theorem is generalized to multidimensional evolution systems and systems which are not translation-invariant. Furthermore, we consider applications of this theorem to the equations of associativity and non-homogeneous systems of hydrodynamic type, and obtain integrable canonical Hamiltonian finite-dimensional reductions of these systems.

Conservation of the Hamiltonian property of a flow, if we restrict this flow to the set of stationary points of a Hamiltonian flow commuting with the first one, was first noted by Novikov in [158] for the restriction of the Korteweg–de Vries equation (KdV) to the set of stationary solutions of the higher KdV. In particular, all finite-gap solutions of the KdV equation are obtained exactly by this construction, namely, by restricting the KdV equation to the set of stationary solutions of corresponding commuting Hamiltonian flows and solving the resulting canonical Hamiltonian integrable finite-dimensional dynamical systems.

In the paper [13] (see also [32], [162]) Bogoyavlenskii and Novikov proved the following remarkable theorem on the connection between Hamiltonian formalisms of stationary and non-stationary problems: from the commutativity of two one-component Hamiltonian flows of the following special form:

$$u_t = \frac{d}{dx}\frac{\delta I}{\delta u(x)}, \qquad I = \int L(u, u_x, \ldots, u_{(n)})\,dx, \qquad (5.65)$$

$$u_t = \frac{d}{dx} \frac{\delta I_1}{\delta u(x)}, \qquad I_1 = \int L_1(u, u_x, \ldots, u_{(k)})\, dx, \qquad (5.66)$$

it follows that for $k < n$ the flow (5.66) is Hamiltonian on the set of stationary points of the flow (5.65), and the Hamiltonian $(-Q_h)$ of the flow (5.66) on the space of solutions of the stationary equation

$$\frac{\delta I}{\delta u(x)} = -h, \qquad h = \text{const},$$

is defined by the following relation:

$$\left(\frac{\delta I}{\delta u(x)} + h \right) \frac{d}{dx} \frac{\delta I_1}{\delta u(x)} \equiv \frac{dQ_h}{dx}.$$

Moreover, in [13] it is conjectured that this theorem is also valid for $k \geq n$ (generally speaking, for the proof, which is presented in [13], the relation $k < n$ is essential) and furthermore that this statement can be generalized to evolution equations that are Hamiltonian with respect to other constant Poisson structures of the form

$$A_{2r+1} = \sum_{i=0}^{r} c_i \frac{d^{2i+1}}{dx^{2i+1}}, \qquad c_i = \text{const}.$$

In [65], [66] Gelfand and Dikii considered the generalization of this theorem to Hamiltonian systems of Lax type

$$L_t = [L, A]$$

(for systems of Lax type, which are Hamiltonian with respect to the Poisson structure of Gelfand and Dikii and which are restricted to the set of stationary points of higher commuting flows from the corresponding hierarchy, we note that the Lagrangian property for the corresponding stationary problem in the case of any ordinary differential operators L and A of coprime orders was established by Veselov in [182]). After that, the problem of generalization of the Bogoyavlenskii–Novikov theorem to evolution systems that are Hamiltonian with respect to an arbitrary constant matrix local Poisson bracket was posed by Gelfand, Manin and Shubin in the paper [71]. Furthermore, the problem on a generalization of the Bogoyavlenskii–Novikov results and the Gelfand–Dikii results to the multidimensional case was posed by Manin (see [97], p. 12). In this connection the question was always about restrictions of Hamiltonian evolution flows with concrete fixed infinite-dimensional Poisson structures. The Hamiltonian property of the flow on a corresponding stationary manifold was induced by the infinite-dimensional Hamiltonian structure of commuting evolution flows, that is, the finite-dimensional Hamiltonian structure was generated (it was a reduction) by the infinite-dimensional structure. All these theorems and conjectures were formulated as statements or assumptions on the connection between the corresponding Hamiltonian formalisms of stationary and non-stationary problems.

In the present author's papers [100], [104] for the first time the Bogoyavlenskii–Novikov principle was formulated and proved for all Poisson brackets. Moreover, it is interesting that, as the present author showed, the condition that the initial non-stationary problem is Hamiltonian is not at all essential. In [100], [104] the present author has formulated and proved the following new and considerably more *general principle*: *any evolution system is a canonical Hamiltonian system on the set of stationary points of its non-degenerate integral.*

We recall that for any functional I the space of solutions of the Lagrangian system

$$\frac{\delta I}{\delta u^i(x)} = 0$$

is called the *stationary manifold* or the *set of stationary points* of this integral.

All the theorems, conjectures, and assumptions mentioned above (including generalizations to the multidimensional case and the case of systems that are not translation-invariant) are direct consequences of this general principle. As Veselov showed subsequently [183] (see also [184], [185]), this general principle is also valid for discrete systems.

We also note that all the preceding papers started with two commuting Hamiltonian flows and, in this case, the invariance of stationary solutions of one of the flows with respect to the other flow, which commutes with the first one, is obvious. This is a direct consequence of the fact that the flows commute. Because of the Hamiltonian property of these commuting evolution flows, the problem always reduces to a restriction of one of the flows to a stationary manifold of a corresponding non-degenerate integral. In our formulation, for an arbitrary evolution flow the invariance of the set of stationary points of its integral is not already evident by itself. Nevertherless, this is also an absolutely general fact. In fact, it is a formally algebraic statement, which does not require any functional assumptions. Here we prove this simple but very important general lemma.

Let us consider an arbitrary multidimensional evolution system of equations

$$u_t^k = F^k(x, u, \ldots, u_{(r)}^s, \ldots), \tag{5.67}$$

where

$$x = (x^1, \ldots, x^n), \qquad u = (u^1, \ldots, u^N), \qquad r = (r_1, \ldots, r_n),$$

$$u_{(r)}^s = \frac{\partial^{|r|} u^s}{\partial(x^1)^{r_1} \cdots \partial(x^n)^{r_n}}, \qquad |r| = r_1 + \cdots + r_n.$$

Let

$$I = \int L(x, u, \ldots, u_{(r)}^s, \ldots) \, d^n x \tag{5.68}$$

be an integral of motion of the system (5.67).

Lemma 5.2 ([85]) *The set of stationary points of the integral I, that is, the space of solutions of the Lagrangian system*

$$\frac{\delta I}{\delta u^k(x)} = 0, \qquad 1 \le k \le N, \tag{5.69}$$

is invariant for the evolution flow (5.67).

Proof. Let $\{u^k(x), \ 1 \le k \le N\}$ be an arbitrary solution of the Lagrangian system (5.69). We consider the vector-function $\{u^k(x,t)\}$, which is the solution of the evolution system (5.67) with the initial conditions $u^k(x,0) = u^k(x)$. Then, for arbitrary constants t_0 and ε and also for an arbitrary vector-function $\{w^k(x)\}$, we consider the vector-function $\{u^k(x,t,\varepsilon)\}$, which is the solution of the evolution system (5.67) with the following initial conditions at $t = t_0$:

$$u^k(x, t_0, \varepsilon) = u^k(x, t_0) + \varepsilon w^k(x).$$

By virtue of the definition of variational derivative, for any functional I there is always the following relation:

$$\frac{d}{d\varepsilon} I[u^k(x,t,\varepsilon)]\big|_{\varepsilon=0} = \int \frac{\delta I}{\delta u^s(x)} [u^k(x,t,0)] v^s(x,t) \, d^n x, \quad (5.70)$$

where

$$v^s(x,t) = \frac{d}{d\varepsilon} u^s(x,t,\varepsilon)\bigg|_{\varepsilon=0}.$$

Since I is an integral of system (5.67), the left-hand side of relation (5.70) does not depend on t and the right-hand side vanishes at $t = 0$, since $u^k(x,t_0,0) = u^k(x,t_0)$ by definition, and hence $u^k(x,t,0) = u^k(x,t)$ as solutions of the evolution system (5.67) with the same initial conditions at $t = t_0$, and $u^k(x,0)$ is a solution of the Lagrangian system (5.69). Thus,

$$\int \frac{\delta I}{\delta u^s(x)} [u^k(x,t,0)] v^s(x,t) \, d^n x = 0. \quad (5.71)$$

At $t = t_0$ the relation (5.71) assumes the form

$$\int \frac{\delta I}{\delta u^s(x)} [u^k(x,t_0)] w^s(x) \, d^n x = 0,$$

whence, since t_0 and $\{w^k(x)\}$ are arbitrary, it immediately follows that $u(x,t)$ is a solution of the Lagrangian system (5.69) for any t, that is, the space of solutions of the Lagrangian system (5.69) is invariant with respect to the evolution flow (5.67).

It is important to note that Lemma 5.2 can be proved in the framework of the formal calculus of variations. In the present author's paper [100] (see also [104]) in 1984 there was published an explicit construction (later we present it in more detail) of the restriction of an evolution flow to the stationary manifold of its nondegenerate integral, for which all statements and formulae were established in the framework of differential algebra and formal calculus of variations (including the statement on invariance formulated in Lemma 5.2, since it automatically follows from the proof of the construction presented). Furthemore, Lemma 5.2 immediately follows from the following identity, which is satisfied for any evolution flow of the form (5.67) and any functional of the form (5.68):

$$\left(\frac{\delta I}{\delta u^i(x)} \right)_t - \frac{\delta}{\delta u^i(x)} (I_t) =$$

$$= (-1)^{|r|+1} \frac{\partial^{|r|}}{\partial(x^1)^{r_1} \cdots \partial(x^n)^{r_n}} \left(\frac{\partial F^s}{\partial u_{(r)}^i} \frac{\delta I}{\delta u^s(x)} \right). \quad (5.72)$$

If the functional $I = \int L(x, u, \ldots, u_{(r)}^s, \ldots) \, d^n x$ in formula (5.72) is an integral of the evolution flow (5.67), then

$$\frac{\delta}{\delta u^i(x)} (I_t) = 0$$

and it immediately follows from formula (5.72) that the space of solutions of the Lagrangian system (5.69) is invariant with respect to the evolution flow (5.67).

To simplify the notation and too cumbersome formulae we consider in detail here only the one-dimensional case ($n = 1$), but then we point out the necessary modification of all key formulae for the multidimensional case and give simple examples of applications of our construction to the multidimensional case (we note that all calculations and results presented below are valid in the multidimensional case).

So we consider an arbitrary one-dimensional evolution system of equations:

$$u_t^k = F^k(x, u, u_x, \ldots, u_{(i)}^j, \ldots), \quad (5.73)$$

$$0 \leq i \leq m_{kj}, \quad 1 \leq k, j \leq N, \quad u_{(i)}^j = \frac{\partial^i u^j}{\partial x^i},$$

possessing a non-degenerate integral of motion

$$I = \int L(x, u, u_x, \ldots, u_{(i)}^j, \ldots) \, dx, \quad (5.74)$$

$$0 \leq i \leq n_j, \quad 1 \leq j \leq N.$$

Then there exists a function $Q(x, u, u_x, \ldots, u_{(i)}^j, \ldots)$ such that

$$\frac{\delta I}{\delta u^k(x)} F^k \equiv \frac{dQ}{dx}. \quad (5.75)$$

The relation (5.75) is equivalent to the condition that the functional (5.74) is an integral of motion of the evolution system (5.73).

By virtue of Lemma 5.2 the space of solutions of the corresponding Lagrangian system (5.69) is invariant for the evolution system (5.73).

Under the condition of non-degeneracy

$$\det\left(\frac{\partial^2 L}{\partial u_{(n_k)}^k \partial u_{(n_j)}^j}\right) \neq 0$$

the Lagrangian L defines the $\left(\sum_{s=1}^{N} 2n_s\right)$-dimensional phase space T of the Lagrangian system (5.69). In the standard way, we introduce the phase variables (q_i^k, p_k^i) for the Lagrangian system (5.69) defined by the functional (5.74):

$$\begin{cases} q_i^k = u_{(i-1)}^k, & 1 \leq i \leq n_k, \quad 1 \leq k \leq N, \\ p_k^i = \sum_{s=0}^{n_k - i} (-1)^s \frac{d^s}{dx^s} \frac{\partial L}{\partial u_{(i+s)}^k}. \end{cases} \tag{5.76}$$

The Hamiltonian of the Lagrangian system (5.69) defined by the functional (5.74) has the form

$$H = \sum_{k=1}^{N}\left(\sum_{s=1}^{n_k-1} p_k^s q_{s+1}^k + p_k^{n_k}(q_{n_k}^k)_x\right) - L\left(x, q_i^k, (q_{n_k}^k)_x\right). \tag{5.77}$$

For the non-degenerate Lagrangian L, from the relations

$$p_k^{n_k} = \frac{\partial L}{\partial (q_{n_k}^k)_x}, \qquad 1 \leq k \leq N,$$

we find an expression for $(q_{n_k}^k)_x$ and hence for the Hamiltonian H via the phase variables (5.76): $(q_{n_k}^k)_x = f(x, q_i^j, p_s^{n_s})$.

On the phase space T the standard finite-dimensional Lagrangian Poisson bracket is defined by

$$\{h, g\} = \sum_{k=1}^{N}\sum_{i=1}^{n_k}\left(\frac{\partial h}{\partial q_i^k}\frac{\partial g}{\partial p_k^i} - \frac{\partial h}{\partial p_k^i}\frac{\partial g}{\partial q_i^k}\right). \tag{5.78}$$

Correspondingly, the Lagrangian system (5.69) assumes the form of a finite-dimensional canonical Hamiltonian dynamical system:

$$\begin{cases} (q_i^k)_x = \frac{\partial H}{\partial p_k^i} \equiv \{q_i^k, H\}, \\ (p_k^i)_x = -\frac{\partial H}{\partial q_i^k} \equiv \{p_k^i, H\}. \end{cases} \tag{5.79}$$

Now if we express the function Q via the phase variables $\widehat{Q} = \widehat{Q}(q, p)$, then the following theorem is valid.

Theorem 5.6 ([100], [104]) *An arbitrary one-dimensional evolutionary flow (5.73) restricted to the set of stationary points of its non-degenerate integral I (5.74) is a canonical Hamiltonian finite-dimensional dynamical system with Hamiltonian* $-\widehat{Q}$:

$$\begin{cases} (q_i^k)_t = -\{q_i^k, \widehat{Q}\}, \\ (p_k^i)_t = -\{p_k^i, \widehat{Q}\}, \end{cases} \tag{5.80}$$

moreover, in the case of translation invariance of the flow (5.73) and the integral I (5.74), the Hamiltonian $-\widehat{Q}$ *is in involution with the Hamiltonian H (5.77) of the Lagrangian system (5.69) defined by the functional (5.74) with respect to the standard Lagrangian finite-dimensional Poisson bracket (5.78) on the phase space T:*

$$\{\widehat{Q}, H\} = 0.$$

Proof. First of all, we note that if we introduce the variables (q_i^k, p_k^i) by formulae (5.76) and define the function H by formula (5.77) (but we shall not consider condition (5.69) for now), then we shall obtain the following obvious but very important functional *identities* on the space of vector-functions $\{u^i(x)\}$:

$$\begin{cases} (p_k^1)_x + \frac{\partial H}{\partial q_1^k} \equiv -\frac{\delta I}{\delta u^k(x)} \, ; \\ (p_k^i)_x + \frac{\partial H}{\partial q_i^k} \equiv 0, \quad i > 1; \\ (q_i^k)_x - \frac{\partial H}{\partial p_k^i} \equiv 0. \end{cases} \tag{5.81}$$

From relations (5.76) and (5.81) all derivatives $u_{(i)}^k$ can be expressed via the variables q_i^k, p_k^i and $(p_k^1)_{(j)}$. We rewrite identity (5.75) in these new variables, substituting the function $Q = Q(x, q_i^k, p_k^i, (p_k^1)_{(j)})$:

$$\left(-(p_k^1)_x - \frac{\partial H}{\partial q_1^k}\right) F^k \equiv \frac{\partial Q}{\partial x} + \sum_{i=1}^{n_k} \frac{\partial Q}{\partial q_i^k} \frac{\partial H}{\partial p_k^i} -$$

$$- \sum_{i=2}^{n_k} \frac{\partial Q}{\partial p_k^i} \frac{\partial H}{\partial q_i^k} + \sum_{j=0}^{s_k} \frac{\partial Q}{\partial (p_k^1)_{(j)}} (p_k^1)_{(j+1)}. \tag{5.82}$$

We introduce the notation $\widehat{G}$ for a function $G(x, q_i^k, p_k^i, (p_k^1)_{(j)})$ expressed via the phase variables with the help of the relations

$$(p_k^1)_x = -\frac{\partial H}{\partial q_1^k}$$

and the second and third identities from formulae (5.81): $\widehat{G} = \widehat{G}(x, q_i^k, p_k^i)$. This corresponds exactly to the reduction to the space of solutions of the Lagrangian system (5.69). From the identity (5.82), in particular, it follows that

$$\frac{\partial \widehat{Q}}{\partial x} + \sum_{i=1}^{n_k} \frac{\partial \widehat{Q}}{\partial q_i^k} \frac{\partial H}{\partial p_k^i} - \sum_{i=1}^{n_k} \frac{\partial \widehat{Q}}{\partial p_k^i} \frac{\partial H}{\partial q_i^k} = 0, \qquad (5.83)$$

that is,

$$\frac{\partial \widehat{Q}}{\partial x} + \{\widehat{Q}, H\} \equiv 0.$$

Correspondingly, in the translation-invariant case both functions $\widehat{Q}$ and H must be integrals in involution for the finite-dimensional canonical Hamiltonian system (5.79):

$$\{\widehat{Q}, H\} \equiv 0.$$

We consider the dynamics of the phase variables (q_i^k, p_k^i) by virtue of the evolution flow (5.73). It is obvious that

$$(q_1^k)_t = \widehat{F}^k.$$

The proof of the following lemma meets the very great difficulties in all of the proposed construction.

Lemma 5.3 *It follows from (5.82) that*

$$\widehat{F}^k = -\frac{\partial \widehat{Q}}{\partial p_k^1}. \qquad (5.84)$$

By virtue of Lemma 5.3

$$(q_1^k)_t = -\frac{\partial \widehat{Q}}{\partial p_k^1}.$$

The dynamics of the rest of the phase variables is generated by the dynamics of this variable q_1^k (in fact, it must be so since $q_1^k = u^k(x)$).

Corollary 5.1 *Any commuting evolution flows $u_t^k = F^k$ and $u_\tau^s = G^s$ possessing a common integral of motion I generate (by the explicit construction presented) integrals Q_1, Q_2 of the stationary system*

$$\frac{\delta I}{\delta u^i(x)} = 0,$$

and

$$\{Q_1, Q_2\} = \text{const}$$

for the standard Lagrangian Poisson bracket on the phase space of the stationary system.

Corollary 5.2 *Let us consider two arbitrary Hamiltonian flows*

$$u_t^k = \{u^k, H_1\} \equiv L^{ks}\frac{\delta H_1}{\delta u^s(x)}, \qquad (5.85)$$

$$u_\tau^k = \{u^k, H_2\} \equiv L^{ks}\frac{\delta H_2}{\delta u^s(x)}, \qquad (5.86)$$

defined with the help of the same arbitrary local Poisson structure

$$\{u^k(x), u^s(y)\} = L^{ks}[u(x)]\delta(x - y),$$

where L^{ks} is a differential Hamiltonian operator whose coefficients depend on x, $u^k(x)$ and their derivatives $u_{(n)}^k(x)$. If the Hamiltonians

$$H_i = \int h_i(x, u, u_x, \ldots)\, dx$$

of the flows (5.85) and (5.86) are in involution, that is,

$$\{H_1, H_2\} = 0,$$

then these evolution flows commute, since the condition of commutativity of the flows has the form

$$L^{ks}\frac{\delta}{\delta u^s(x)}\{H_1, H_2\} = 0,$$

and Theorem 5.6 establishes the canonical Hamiltonian property of the evolution flow (5.85) on the set of stationary points of the flow (5.86):

$$\frac{\delta H_2}{\delta u^s(x)} = \frac{\delta g}{\delta u^s(x)}, \qquad \frac{\delta g}{\delta u^s(x)} \in \text{Ker}(L^{ks}).$$

In particular, if an evolution flow is defined by a compatible pair of Poisson brackets, then it has higher analogues, which satisfy the conditions of Corollary 5.2 (for example, systems of Lax type $L_t = [L, A]$ and many known integrable systems of partial differential equations possess these properties). Many well-known partial solutions of evolution systems (the finite-gap solutions of the Korteweg–de Vries equation, the Langmuir solitons, and many others) can be obtained by the reduction of the system to the set of stationary points of the corresponding integral according to the general proposed scheme.

The proof of the theorem is completely generalized to the multi-dimensional case. The proposed natural reduction gives the possibility of reducing the dimension of a system that has a local conservation law, and the reduced system always possesses the canonical Hamiltonian structure. We give the corresponding multidimensional formulae and consider their application to concrete model examples. In the case of degenerate conservation laws it is necessary to apply the Dirac scheme, and the theorem can be also generalized to this case.

Thus we consider an arbitrary multidimensional evolution system of equations (5.67). Let

$$I = \int L(x, u, \ldots, u^s_{(r)}, \ldots) \, d^n x \qquad (5.87)$$

be an integral of motion of the evolution system (5.67), that is, there exist functions

$$Q^m = Q^m(x, u, \ldots, u^s_{(r)}, \ldots), \quad 1 \le m \le n,$$

such that the following relation is satisfied:

$$\frac{\delta I}{\delta u^k(x)} F^k \equiv \sum_{m=1}^{n} D_m Q^m,$$

where D_m is the operator of total derivative with respect to the variable x^m:

$$D_m = \frac{\partial}{\partial x^m} + u^k_{x^m} \frac{\partial}{\partial u^k} + \cdots.$$

We assume that the highest derivatives of the field variables $u^k(x)$ with respect to x^1 that occur in the Lagrangian L are not mixed and have order n_k: $u^k_{(n_k)(0)\cdots(0)}$. In this case the phase variables of the stationary system

$$\frac{\delta I}{\delta u^k(x)} = 0 \qquad (5.88)$$

are defined by the following formulae (as in the one-dimensional case, we also require the non-degeneracy of the Lagrangian L in the corresponding highest derivatives $u^k_{(n_k)(0)\cdots(0)}$ with respect to the variable x^1):

$$\begin{cases} q^k_s = u^k_{(s-1)(0)\cdots(0)}, & 1 \le s \le n_k, \quad 1 \le k \le N, \\ p^s_k = \sum_{j=0}^{n_k-s} (-1)^j D^j_1 \frac{\delta \tilde{I}}{\delta u^k_{(s+j)(0)\cdots(0)}}, \end{cases} \qquad (5.89)$$

where

$$\tilde{I} = \int L(x, u, \ldots, u^s_{(r)}, \ldots)\, dx^2 \cdots dx^n. \qquad (5.90)$$

In formulae (5.89), when we calculate variational derivatives, all the derivatives of the field variables $u^k(x)$ of different orders with respect to x^1 are considered as independent variables (in fact, this must be so for functionals of the form (5.90)). Correspondingly, the Hamiltonian of the stationary system has the form

$$H = \int (p^j_k q^k_{j+1} + p^{n_k}_k (q^k_{n_k})_{x^1} - L)\, dx^2 \cdots dx^n,$$

and the Lagrangian Poisson bracket on the phase space is defined by

$$\{H, G\} = \sum_{k=1}^{N} \sum_{i=1}^{n_k} \int \left(\frac{\delta H}{\delta q^k_i} \frac{\delta G}{\delta p^i_k} - \frac{\delta H}{\delta p^i_k} \frac{\delta G}{\delta q^k_i} \right) dx^2 \cdots dx^n. \qquad (5.91)$$

Correspondingly, the Lagrangian system (5.88) assumes the form of the canonical Hamiltonian dynamical system:

$$\begin{cases} (q^k_i)_{x^1} = \frac{\delta H}{\delta p^i_k} \equiv \{q^k_i, H\}, \\ (p^i_k)_{x^1} = -\frac{\delta H}{\delta q^k_i} \equiv \{p^i_k, H\}. \end{cases} \qquad (5.92)$$

In just the same way as in the one-dimensional case, after introducing the phase variables by formulae (5.89), we obtain the following identities on the space of vector-functions $\{u^k(x)\}$:

$$\begin{cases} (p_k^1)_{x^1} + \frac{\delta H}{\delta q_1^k} \equiv -\frac{\delta I}{\delta u^k}\,; \\ (p_k^i)_{x^1} + \frac{\delta H}{\delta q_i^k} \equiv 0,\ \ i > 1; \\ (q_i^k)_{x^1} - \frac{\delta H}{\delta p_k^i} \equiv 0. \end{cases} \qquad (5.93)$$

Theorem 5.7 ([100], [104]) *Any multidimensional evolution flow (5.67) on the set of stationary points of its integral (5.87) is a canonical Hamiltonian system with Hamiltonian*

$$G = -\int \widehat{Q}^1\, dx^2 \cdots dx^n\ :$$

$$\begin{cases} (q_i^k)_t = \frac{\delta G}{\delta p_k^i} \equiv \{q_i^k, G\}, \\ (p_k^i)_t = -\frac{\delta G}{\delta q_i^k} \equiv \{p_k^i, G\}, \end{cases} \qquad (5.94)$$

and

$$\frac{\partial G}{\partial x^1} + \{G, H\} = 0.$$

Example 5.1 We illustrate the proposed scheme in the simplest multidimensional situation when it requires the minimum of necessary calculations. We consider the linear $(2 + 1)$-dimensional flow

$$u_t = u_x + u_y$$

and its non-degenerate quadratic integral

$$I = \frac{1}{2}\int (u_x^2 + a u_y^2 + b u_x u_y)\, dx\, dy.$$

According to the general scheme we have

$$\frac{\delta I}{\delta u} u_t = (Q^1)_x + (Q^2)_y,$$

$$Q^1 = -\frac{1}{2}u_x^2 - u_x u_y - \frac{1}{2}(a - b)u_y^2.$$

According to (5.89), we introduce the phase variables q, p of the stationary problem $\frac{\delta I}{\delta u} = 0$:

$$q = u, \qquad p = u_x + \frac{1}{2} b u_y.$$

Then

$$q_t = p + \left(1 - \frac{b}{2}\right) q_y = \frac{\delta G}{\delta p},$$

$$p_t = \left(1 - \frac{b}{2}\right) p_y + \left(\frac{b^2}{4} - a\right) q_{yy} = -\frac{\delta G}{\delta q},$$

where

$$G = -\int Q^1 \, dy = \frac{1}{2} \int \left[p^2 + (2 - b) p q_y + \left(\frac{1}{4} b^2 - a\right) q_y^2\right] dy.$$

The Hamiltonian H of the stationary Lagrangian system $\frac{\delta I}{\delta u} = 0$ has the form

$$H = \int \left[\frac{1}{2} p^2 - \frac{1}{2} b p q_y + \left(\frac{1}{8} b^2 - \frac{1}{2} a\right) q_y^2\right] dy,$$

$$\{H, G\} = 0.$$

Example 5.2 We consider now the simplest non-linear evolution flow

$$u_t = u_x^2 - \varepsilon u_y^2, \tag{5.95}$$

possessing the non-degenerate quadratic integral

$$I = -\frac{1}{2} \int (u_x^2 + \varepsilon u_y^2) \, dx \, dy, \tag{5.96}$$

$$\frac{\delta I}{\delta u} = u_{xx} + \varepsilon u_{yy}, \tag{5.97}$$

$$\frac{\delta I}{\delta u} u_t = \left(\frac{u_x^3}{3} - \varepsilon u_y^2 u_x\right)_x - \left(\varepsilon^2 \frac{u_y^3}{3} - \varepsilon u_x^2 u_y\right)_y,$$

$$Q^1 = \frac{u_x^3}{3} - \varepsilon u_y^2 u_x.$$

We introduce the phase variables $q = u$, $p = -u_x$. The reduction of the $(2+1)$-dimensional equation (5.95) to the set of stationary points of its integral (5.96) has the form of a canonical Hamiltonian $(1+1)$-dimensional evolution system:

$$q_t = p^2 - \varepsilon q_y^2 = \frac{\delta G}{\delta p},$$

$$p_t = -2\varepsilon p q_{yy} - 2\varepsilon q_y p_y = -\frac{\delta G}{\delta q}, \qquad (5.98)$$

where

$$G = -\int \widehat{Q}^1 \, dy = \int \left(\frac{p^3}{3} - \varepsilon p q_y^2 \right) dy.$$

The Hamiltonian of the stationary Lagrangian system (5.97) has the form

$$H = \frac{1}{2} \int (-p^2 + \varepsilon q_y^2) \, dy,$$

$$\{H, G\} = 0.$$

The system (5.98) is equivalent to the non-linear equation

$$q_{tt} = -4\varepsilon (q_t q_y)_y - 8\varepsilon^2 q_y^2 q_{yy}.$$

Now we consider examples of applications of Theorem 5.6.

Example 5.3 Let us consider the Gibbons–Tsarev system, that is, the non-homogeneous two-component system of hydrodynamic type (see [72], [180])

$$u_t = v u_x - \frac{1}{u - v},$$

$$v_t = u v_x + \frac{1}{u - v}. \qquad (5.99)$$

The system (5.99), as Gibbons and Tsarev showed (see [72], [180]), describes all two-parametric reductions of the Benney equations, that is, the system of equations for two-dimensional flows of a non-viscous incompressible fluid with free surface in the field of gravity in the approximation of long waves ([8]):

$$u_t + u u_x - u_y \int_0^y u_x \, dy + h_x = 0,$$

$$h_t + \left(\int_0^h u \, dy \right)_x = 0, \qquad (5.100)$$

where $-\infty < x < \infty$ is the horizontal coordinate, $0 \leq y$ is the vertical coordinate, t is time, $u = u(x, y, t)$ is the horizontal component of the velocity, and $h(x, t)$ is the height of the free surface (the normalization is the following: the acceleration of gravity and the density of fluid are equal to 1). The vertical component of the velocity $v(x, y, t)$ is eliminated from the system of Benney equations (5.100) by virtue of the equation of continuity $u_x + v_y = 0$ and the boundary condition $v = 0$ at $y = 0$. Benney also showed that the system (5.100) implies the following homogeneous system of hydrodynamic type with infinitely many components (the so-called moment equations):

$$A_t^k + A_x^{k+1} + kA^{k-1}A_x^0 = 0, \qquad k = 0, 1, 2, \ldots, \qquad (5.101)$$

where

$$A^k(x, t) = \int_0^h u^k(x, y, t)\, dy, \qquad k \geq 0,$$

are the corresponding moments. There are the well-known $2N$-parametric reductions of Zakharov [194] for the Benney equations

$$\begin{cases} \eta_t^i + (q^i \eta^i)_x = 0, \\ q_t^i + q^i q_x^i + P_x = 0, \end{cases} \qquad (5.102)$$

$$P = \sum_{k=1}^N \eta^k, \qquad i = 1, \ldots, N,$$

obtained from the moment equations (5.101) for

$$A^k = \sum_{s=1}^N \eta^s (q^s)^k.$$

The system of equations (5.102) models the flows of an N-layer incompressible fluid: $\eta^i(x, t)$ is the thickness of the ith layer, and $q^i(x, t)$ is its velocity. The Benney system (5.102) is a homogeneous diagonalizable system of hydrodynamic type, which possesses three local Poisson structures of hydrodynamic type and is integrated by the generalized hodograph method [171], [180]. Tsarev has built a complete family of Combescure transformations for the Egorov coordinate system corresponding to the N-layer

Benney system (5.102). So one can find symmetries of the system and construct solutions of the system with the help of generalized hodograph formulae (see [180]). Along with the reduction of Zakharov (5.102), there are other known reductions of the moment equations (5.101), and these reductions also result in diagonalizable integrable systems of hydrodynamic type (see [171]). Gibbons and Tsarev [72] have studied the very interesting problem of describing all possible finite-parametric reductions of the moment equations (5.101), that is, N-dimensional submanifolds $A^k = A^k(u^1, \ldots, u^N)$ that are invariant with respect to (5.101) (here $u^i(x, t)$ are new field variables). It is shown that one always can take the variables $A^0, A^1, \ldots, A^{N-1}$ as the new variables u^i and, correspondingly, it is sufficient to look for reductions of the form

$$
\begin{aligned}
A^N &= f(A^0, \ldots, A^{N-1}), \\
A^{N+1} &= A^{N+1}(A^0, \ldots, A^{N-1}), \\
&\cdots\cdots\cdots\cdots\cdots\cdots\cdots\cdots\cdots\cdots\cdots \\
A^{N+k} &= A^{N+k}(A^0, \ldots, A^{N-1}), \\
&\cdots\cdots\cdots\cdots\cdots\cdots\cdots\cdots\cdots\cdots\cdots
\end{aligned}
\tag{5.103}
$$

The reduction (5.103) is possible if and only if the following system of equations for the function f is satisfied ([72], [180]):

$$
\begin{aligned}
&\frac{\partial}{\partial A^r}\left(\frac{\partial f}{\partial A^{N-1}}\frac{\partial f}{\partial A^s} + \frac{\partial f}{\partial A^{s-1}}\right) = \\
&= \frac{\partial}{\partial A^s}\left(\frac{\partial f}{\partial A^{N-1}}\frac{\partial f}{\partial A^r} + \frac{\partial f}{\partial A^{r-1}}\right), \quad 1 \leq s, r \leq N-1, \\
&\frac{\partial}{\partial A^0}\left(\frac{\partial f}{\partial A^{N-1}}\frac{\partial f}{\partial A^s} + \frac{\partial f}{\partial A^{s-1}}\right) = \\
&= \frac{\partial}{\partial A^s}\left[-NA^{N-1} + \frac{\partial f}{\partial A^{N-1}}\left((N-1)A^{N-2} + \right.\right. \\
&\left.\left. + \frac{\partial f}{\partial A^0}\right) + \sum_{p=0}^{N-2} pA^{p-1}\frac{\partial f}{\partial A^p}\right].
\end{aligned}
\tag{5.104}
$$

As is shown in [72], [180], all the reductions are diagonalizable integrable homogeneous systems of hydrodynamic type. The system (5.104) is very interesting and non-trivial even in the case of two-parametric reductions. In this case

$$\frac{\partial}{\partial A^0}\left[\left(\frac{\partial f}{\partial A^1}\right)^2 + \frac{\partial f}{\partial A^0}\right] = \frac{\partial}{\partial A^1}\left[-2A^1 + \frac{\partial f}{\partial A^1}\left(A^0 + \frac{\partial f}{\partial A^0}\right)\right].$$

After introducing new variables

$$t = A^0, \quad x = A^1, \quad z = f(A^0, A^1) + (A^0)^2/2,$$

we obtain the non-homogeneous quasilinear equation

$$z_{tt} + z_x z_{xt} - z_t z_{xx} + 1 = 0,$$

which is equivalent to the non-homogeneous system of hydrodynamic type ($a = z_t$, $b = z_x$):

$$\begin{cases} a_t = ab_x - ba_x - 1, \\ b_t = a_x. \end{cases}$$

After the change of variables

$$u = \frac{1}{2}(-b + \sqrt{b^2 + 4a}), \quad v = \frac{1}{2}(-b - \sqrt{b^2 + 4a}),$$

the Gibbons–Tsarev system assumes the form (5.99). The system (5.99) is integrable (see [73]) but no Hamiltonian representation for system (5.99) has been found. Meanwhile it is easy to find non-degenerate quadratic laws for the system (5.99) (see [62]):

$$I = \int L\,dx = \int [(v - u)(u_x^2 - v_x^2) + c(u + v)]\,dx, \qquad (5.105)$$

where c is an arbitrary constant (for the system (5.99), this is the general form of the conservation law that is quadratic in first derivatives).

The application of the explicit scheme of Theorem 5.6 to the conservation law automatically leads to an integrable canonical Hamiltonian finite-dimensional dynamical system. We have:

$$\frac{\delta I}{\delta u} = -u_x^2 + v_x^2 + c - 2[(v - u)u_x]_x,$$

$$\frac{\delta I}{\delta v} = u_x^2 - v_x^2 + c + 2[(v - u)v_x]_x,$$

$$\frac{\delta I}{\delta u}u_t + \frac{\delta I}{\delta v}v_t = [cuv - 2u_x - 2v_x + (v-u)uv_x^2 - (v-u)vu_x^2]_x,$$

$$Q = cuv - 2u_x - 2v_x + (v-u)uv_x^2 - (v-u)vu_x^2.$$

For the stationary Lagrangian problem

$$\frac{\delta I}{\delta u(x)} = 0, \qquad \frac{\delta I}{\delta v(x) = 0} \qquad (5.106)$$

we introduce the phase variables (q^i, p^i):

$$q^1 = u, \qquad q^2 = v,$$

$$p^1 = \frac{\partial L}{\partial u_x} = 2(v-u)u_x, \qquad p^2 = \frac{\partial L}{\partial v_x} = -2(v-u)v_x.$$

Correspondingly, we obtain

$$u_x = \frac{p^1}{2(q^2 - q^1)}, \qquad v_x = \frac{p^2}{2(q^1 - q^2)},$$

$$\widehat{Q} = cq^1q^2 + \frac{p^2 - p^1}{q^2 - q^1} + \frac{q^1(p^2)^2 - q^2(p^1)^2}{4(q^2 - q^1)}. \qquad (5.107)$$

The reduction of the Gibbons–Tsarev system (5.99) to the stationary manifold of the integral (5.105) has the form:

$$q^1_t = -\{q^1, \widehat{Q}\} = \frac{1}{q^2 - q^1} + \frac{p^1q^2}{2(q^2 - q^1)},$$

$$q^2_t = -\{q^2, \widehat{Q}\} = -\frac{1}{q^2 - q^1} - \frac{p^2q^1}{2(q^2 - q^1)},$$

$$p^1_t = -\{p^1, \widehat{Q}\} = cq^2 + \frac{p^2 - p^1}{(q^2 - q^1)^2} +$$

$$+\frac{q^1(p^2)^2 - q^2(p^1)^2}{4(q^2 - q^1)^2} - \frac{(p^2)^2}{4(q^2 - q^1)},$$

$$p^2_t = -\{p^2, \widehat{Q}\} = cq^1 - \frac{p^2 - p^1}{(q^2 - q^1)^2} -$$

$$-\frac{q^1(p^2)^2 - q^2(p^1)^2}{4(q^2 - q^1)^2} - \frac{(p^1)^2}{4(q^2 - q^1)}. \qquad (5.108)$$

The Hamiltonian H of the stationary Lagrangian problem (5.106) has the form

$$H = p^1 u_x + p^2 v_x - L = \frac{(p^1)^2 - (p^2)^2}{4(q^2 - q^1)} . \qquad (5.109)$$

Since the functions $\widehat{Q}$ and H are independent and in involution $\{\widehat{Q}, H\} = 0$, by virtue of Theorem 5.6 (now it is easy to verify this fact by direct calculation), the canonical Hamiltonian system (5.108) and also the canonical Hamiltonian four-component system corresponding to the stationary Lagrangian problem (5.106) are completely integrable.

In just the same way Theorem 5.6 can also be applied to other homogeneous and non-homogeneous systems of hydrodynamic type that possess non-degenerate higher conservation laws (a classification and description of systems possessing higher conservation laws is a separate problem, which has been quite well developed recently and there is a very rich family of corresponding examples).

As is obvious from formulae (5.107) and (5.109), the functions $\widehat{Q}$ and H, which are in involution with respect to the standard Lagrangian Poisson bracket, are quadratic in the momenta p^1, p^2. Recently Ferapontov and Fordy have established [48], [49], that if we apply Theorem 5.6 to arbitrary non-homogeneous systems of hydrodynamic type possessing a quadratic first-order integral, then we obtain all pairs of functions that are quadratic in momenta and in involution. An effective description of such pairs of functions is an independent very interesting problem, which is important for theoretical mechanics. This problem has not yet been solved in the general case.

Using the bi-Hamiltonian representation of the equation of associativity (5.2), by the Lenard–Magri scheme we find its higher first-order integrals (we shall follow here to the paper [50]). It is convenient to make all necessary calculations in the variables u^1, u^2, u^3, which are connected with the variables a, b, c by the Vieta formulae, since in these variables the first Poisson structure M_1 becomes the constant Poisson structure $\widehat{M}_1$ (see formulae (5.10) and (5.11) in Section 5.1). The second Poisson structure M_2 (see formula (5.26)) in the variables u^i assumes the form

$$\widehat{M}_2 = J M_2 J^T, \qquad (5.110)$$

where the coefficients of the Poisson structure M_2, which depend on the variables a, b, c, must be expressed with the help of the Vieta formulae via the variables u^i, and J is the matrix that is inverse to the Jacobi matrix of the change from the variables a, b, c to the variables u^i:

$$J = \left(\frac{\partial(a, b, c)}{\partial(u^1, u^2, u^3)} \right)^{-1},$$

that is,

$$J = \begin{pmatrix} \frac{(u^1)^2}{(u^1-u^2)(u^1-u^3)} & \frac{2u^1}{(u^1-u^2)(u^1-u^3)} & \frac{1}{(u^1-u^2)(u^1-u^3)} \\ \frac{(u^2)^2}{(u^2-u^1)(u^2-u^3)} & \frac{2u^2}{(u^2-u^1)(u^2-u^3)} & \frac{1}{(u^2-u^1)(u^2-u^3)} \\ \frac{(u^3)^2}{(u^3-u^1)(u^3-u^2)} & \frac{2u^3}{(u^3-u^1)(u^3-u^2)} & \frac{1}{(u^3-u^1)(u^3-u^2)} \end{pmatrix}.$$
$$\tag{5.111}$$

We apply the Lenard–Magri scheme to the Casimir functionals

$$K^m = \int u^m \, dx$$

of the first Poisson structure (5.11) of the equation of associativity. Higher integrals of motion I^m can be found from the relations

$$\widehat{M_1^{ij}} \frac{\delta I^m}{\delta u^j(x)} \equiv \widehat{M_2^{ij}} \frac{\delta K^m}{\delta u^j(x)}, \qquad m = 1, 2, 3. \tag{5.112}$$

Direct calculations result in the following explicit formulae for the higher integrals of motion ($\mathcal{I}^m$ are the densities of the functionals I^m):

$$I^1 = \int \mathcal{I}^1 \, dx = \int \left(\frac{2u^1 - u^2 - u^3}{2(u^2 - u^1)^3(u^3 - u^1)^3} \times \right. \tag{5.113}$$
$$\times \left\{ (u^3 - u^2)^2 (u_x^1)^2 + [(u^2 u^3)_x - u^1(u^2 + u^3)_x]^2 \right\} +$$
$$\left. + \frac{(u^2 - u^1)^2 + (u^3 - u^1)^2}{(u^2 - u^1)^3(u^3 - u^1)^3} \, u_x^1 [(u^2 u^3)_x - u^1(u^2 + u^3)_x] \right) dx,$$

and the integrals I^2 and I^3 are obtained from (5.113) by the change of indices $1 \leftrightarrow 2$ and $1 \leftrightarrow 3$ respectively.

It is easy to verify that by virtue of (5.10) we have

$$\mathcal{I}_t^1 = \mathcal{F}_x^1, \tag{5.114}$$

where

$$\mathcal{F}^1 = \frac{(u^2 + u^3 - u^1)^2 - u^2 u^3}{2(u^2 - u^1)^3(u^3 - u^1)^3} \times \tag{5.115}$$

$$\times \left\{ (u^3 - u^2)^2 u_x^{1\,2} + [(u^2 u^3)_x - u^1(u^2 + u^3)_x]^2 \right\} +$$

$$+ \frac{u^2(u^2 - u^1)^2 + u^3(u^3 - u^1)^2}{(u^2 - u^1)^3(u^3 - u^1)^3} u_x^1 [(u^2 u^3)_x - u^1(u^2 + u^3)_x].$$

The following evident relation is valid:

$$\widehat{M}_1^{ij} \frac{\delta(I^1 + I^2 + I^3)}{\delta u^j(x)} \equiv \widehat{M}_2^{ij} \frac{\delta(K^1 + K^2 + K^3)}{\delta u^j(x)} = 0, \tag{5.116}$$

since the functional

$$K^1 + K^2 + K^3 = \int a \, dx$$

is a Casimir functional of the Poisson structure M_2^{ij}. In fact, it is easy to show that

$$I^1 + I^2 + I^3 = 0,$$

and hence there are only two linearly independent integrals among I^m.

Higher third-order flows of the hierarchy of the equations of associativity (5.10) have the form

$$u^i_{\tau_m} = \widehat{M}_1^{ij} \frac{\delta I^m}{\delta u^j(x)} = \widehat{M}_2^{ij} \frac{\delta K^m}{\delta u^j(x)}, \qquad m = 1, 2, 3. \tag{5.117}$$

The constructed independent higher integrals I^1 and I^2 are quadratic in the first derivatives, that is, they have the form

$$I^1 = \int g_{ij}^1(u) u_x^i u_x^j \, dx, \qquad I^2 = \int g_{ij}^2(u) u_x^i u_x^j \, dx. \tag{5.118}$$

In addition, the system of equations of associativity (5.10) possesses five integrals of motion of hydrodynamic type

$$K^i = \int u^i \, dx, \qquad i = 1, 2, 3, \tag{5.119}$$

$$K^4 = \int \left(u^1 u^2 + u^1 u^3 + u^2 u^3 \right) dx, \qquad (5.120)$$

$$K^5 = \int u^1 u^2 u^3 \, dx. \qquad (5.121)$$

An arbitrary linear combination of the integrals of motion (5.118)–(5.121)

$$I = \int \left(g_{ij}(u) \, u_x^i \, u_x^j + V(u) \right) dx \qquad (5.122)$$

defines a Lagrangian of the motion of a particle in curved space with the metric

$$g_{ij} = \lambda_1 g_{ij}^1 + \lambda_2 g_{ij}^2 \qquad (5.123)$$

under the influence of the potential $V(u)$ that is a cubic polynomial in the variables u^i. Here λ_1 and λ_2 are arbitrary constants. For any choice of these constants with the exception of the following special cases: 1) $\lambda_1 = \lambda_2$, 2) $\lambda_1 = 0$, 3) $\lambda_2 = 0$, the metric (5.123) is non-degenerate, and hence the Lagrangian (5.122) is also non-degenerate.

Thus according to Theorem 5.6, we can consider the restriction of the equation of associativity represented in the form of the system of hydrodynamic type (5.10) to the set of stationary points of the integral (5.122). And then we obtain an integrable finite-dimensional canonical Hamiltonian dynamical system as the reduction of the equation of associativity.

Acknowledgments

This work was partially supported by the Alexander von Humboldt Foundation (Germany), the Russian Foundation for Basic Research (grants Nos. 96-01-01623, 99-01-00010, 02-01-00803, and 03-01-00782), the program of support for the leading scientific schools (grants Nos. 96-15-96027 and 2185.2003.1) and INTAS (grants Nos. 96-0770 and 99-1782).

References

[1] Alekseev V. L. *On non-local Hamiltonian operators of hydro-dynamic type connected with Whitham's equations.* Uspekhi Mat. Nauk. 1995. V. 50. No. 6. P. 165–166; English transl. in Russian Math. Surveys. 1995. V. 50. No. 6. P. 1253–1255.

[2] Antonowicz M., Fordy A. P. *A family of completely integrable multi-Hamiltonian systems.* Phys. Lett. A. 1987. V. 122. P. 95–99.

[3] Antonowicz M., Fordy A. P. *Coupled KdV equations with multi-Hamiltonian structures.* Physica D. 1987. V. 28. P. 345–357.

[4] Antonowicz M., Fordy A. P. *Coupled KdV equations associated with a novel Schrödinger spectral problem.* In: Nonlinear Evolutions. Proceedings of the 4th Workshop on Nonlinear Evolution Equations and Dynamical Systems (Balarac-les-Bains, France, 1987). Ed. J. P. Leon . World Sci. Publishing. Teaneck, NJ. 1988. P. 145–159.

[5] Astashov A. M. *Normal forms of Hamiltonian operators in field theory.* Dokl. Akad. Nauk SSSR. 1983. V. 270. No. 5. P. 363–368; English transl. in Soviet Math. Dokl. 1983. V. 27.

[6] Astashov A. M., Vinogradov A. M. *On the structure of Hamiltonian operators in field theory.* J. Geom. Phys. 1986. V. 3. No. 2. P. 263–287.

[7] Balinskii A. A., Novikov S. P. *Poisson brackets of hydrodynamic type, Frobenius algebras and Lie algebras.* Dokl. Akad. Nauk SSSR. 1985. V. 283. No. 5. P. 1036–1039; English transl. in Soviet Math. Dokl. 1985. V. 32. P. 228–231.

[8] Benney D. J. *Some properties of long non-linear waves.* Stud. Appl. Math. 1973. V. 52. No. 1. P. 45–50.

[9] Besse A. *Manifolds all of whose geodesics are closed.* Springer-Verlag. Berlin–New York. 1978.

[10] Bochner S. *Curvature and Betti numbers.* Ann. of Math. (2). 1948. V. 49. No. 2. P. 379–390.

[11] Bochner S. *Curvature and Betti numbers. II.* Ann. of Math. (2). 1949. V. 50. No. 1. P. 77–93.

[12] Bochner S., Yano K. *Tensor-fields in non-symmetric connections.* Ann. of Math. (2). 1952. V. 56. No. 3. P. 504–519.

[13] Bogoyavlenskii O. I., Novikov S. P. *On the connection between the Hamiltonian formalisms of stationary and non-stationary problems.* Funktsional. Analiz i ego Prilozhen. 1976. V. 10. No. 1. P. 9–13; English transl. in Functional Anal. Appl. 1976. V. 10. No. 1.

[14] Calogero F. *Solutions of certain integrable non-linear PDE's describing nonresonant N-wave interactions.* J. Math. Phys. 1989. V. 30. No. 3. P. 639–654.

[15] Cooke D. B. *Classification results and the Darboux theorem for low-order Hamiltonian operators.* J. Math. Phys. 1991. V. 32. No. 1. P. 109–119.

[16] Cooke D. B. *Compatibility conditions for Hamiltonian pairs.* J. Math. Phys. 1991. V. 32. No. 11. P. 3071–3076.

[17] Dijkgraaf R., Verlinde H., Verlinde E. *Topological strings in d < 1.* Nucl. Phys. B. 1991. V. 352. P. 59–86.

[18] Dirac P. A. M. *Lectures on Quantum Mechanics.* Yeshiva Univ. New York. 1964.

[19] Dorfman I. Ya. *Dirac structures of integrable evolution equations.* Phys. Lett. A. 1987. V. 125. No. 5. P. 240–246.

[20] Dorfman I. Ya. *Dirac structures and integrability of non-linear evolution equations.* Wiley. Chichester. 1993.

[21] Dorfman I. Ya. *On differential operators that generate Hamiltonian structures.* Phys. Lett. A. 1989. V. 140. Nos. 7, 8. P. 378–382.

[22] Dorfman I. Ya. *The Krichever–Novikov equation and local symplectic structures.* Dokl. Akad. Nauk SSSR. 1988. V. 302. No. 4. P. 792–795; English transl. in Soviet Math. Dokl. 1989. V. 38.

[23] Dorfman I. Ya. *Dirac structures of integrable evolution equations.* D.Sc. Dissertation (Phys. and Math.). Leningrad. Steklov Math. Institute. 1989 [in Russian].

[24] Dorfman I. Ya., Mokhov O. I. *Local symplectic operators and structures related to them.* J. Math. Phys. 1991. V. 32. No. 12. P. 3288–3296.

[25] Dorfman I. Ya., Nijhoff F. W. *On a (2+1)-dimensional version of the Krichever–Novikov equation.* Phys. Lett. A. 1991. V. 157. P. 107–112.

[26] Doyle P. W. *Differential-geometric Poisson bivectors in one space variable.* J. Math. Phys. 1993. V. 34. No. 4. P. 1314–1338.

[27] Dubrovin B. A. *Integrable systems and classification of 2-dimensional topological field theories.* Preprint SISSA–162/92/FM, SISSA, Trieste, Italy, 1992; hep-th/9209040.

[28] Dubrovin B. A. *Differential geometry of the space of orbits of a Coxeter group.* Preprint SISSA–29/93/FM, SISSA, Trieste, Italy, 1993; hep-th/9303152.

[29] Dubrovin B. A. *Geometry of 2D topological field theories.* Preprint SISSA–89/94/FM, SISSA, Trieste, Italy, 1994; Lecture Notes in Math. 1996. V. 1620. P. 120–348; hep-th/9407018.

[30] Dubrovin B. A. *Integrable systems in topological field theory.* Nuclear Phys. B. 1992. V. 379. P. 627–689.

[31] Dubrovin B. A. *Flat pencils of metrics and Frobenius manifolds.* Preprint SISSA–25/98/FM, SISSA, Trieste, Italy, 1998; math.dg/9803106.

[32] Dubrovin B. A., Matveev V. B., Novikov S. P. *Non-linear equations of Korteweg–de Vries type, finite zone linear operators, and Abelian varieties.* Uspekhi Mat. Nauk. 1976. V. 31. No. 1. P. 55–136; English transl. in Russian Math. Surveys. 1976. V. 31. No. 1. P. 59–146.

[33] Dubrovin B. A., Novikov S. P. *Hydrodynamics of soliton lattices.* Soviet Sci. Rev. Sect. C. Math. Phys. Rev. 1993. V. 9. No. 4. P. 1–136.

[34] Dubrovin B. A., Novikov S. P. *The Hamiltonian formalism of one-dimensional systems of hydrodynamic type and the Bogolyubov–Whitham averaging method.* Dokl. Akad. Nauk SSSR. 1983. V. 270. No. 4. P. 781–785; English transl. in Soviet Math. Dokl. 1983. V. 27. P. 665–669.

[35] Dubrovin B. A., Novikov S. P. *On Poisson brackets of hydrodynamic type.* Dokl. Akad. Nauk SSSR. 1984. V. 279. No. 2. P. 294–297; English transl. in Soviet Math. Dokl. 1984. V. 30. P. 651–654.

[36] Dubrovin B. A., Novikov S. P. *Hydrodynamics of weakly deformed soliton lattices. Differential geometry and Hamiltonian theory.* Uspekhi Mat. Nauk. 1989. V. 44. No. 6. P. 29–98; English transl. in Russian Math. Surveys. 1989. V. 44. No. 6. P. 35–124.

[37] Dubrovin B. A., Novikov S. P., Fomenko A. T. *Modern geometry: methods and applications, 2nd ed.* Moscow. Nauka. 1986; English transl. of 1st ed.: Parts I, II. Springer-Verlag. New York–Berlin. 1984, 1985.

[38] Ferapontov E. V. *Hamiltonian systems of hydrodynamic type and their realization on hypersurfaces of a pseudo-Euclidean space.* In: Problems of geometry. 1990. V. 22. Moscow. VINITI. P. 59–96; English transl. in J. Soviet Math. 1991. V. 55. P. 1970–1995.

[39] Ferapontov E. V. *Differential geometry of nonlocal Hamiltonian operators of hydrodynamic type.* Funktsional. Analiz i ego Prilozhen. 1991. V. 25. No. 3. P. 37–49; English transl. in Functional Anal. Appl. 1991. V. 25. No. 3. P. 195–204.

[40] Ferapontov E. V. *Nonlocal matrix Hamiltonian operators. Differential geometry and applications.* Teoret. Mat. Fiz. 1992. V. 91. No. 3. P. 452–462; English transl. in Theoret. and Math. Phys. 1992. V. 91.

[41] Ferapontov E. V. *Dirac reduction of the Hamiltonian operator* $\delta^{IJ}\frac{d}{dx}$ *to a submanifold of the Euclidean space with flat normal connection.* Funktsional. Analiz i ego Prilozhen. 1992. V. 26. No. 4. P. 83–86; English transl. in Functional Anal. Appl. 1992. V. 26. No. 4. P. 298–300.

[42] Ferapontov E. V. *On integrability of* 3×3 *semi-Hamiltonian hydrodynamic type systems which do not possess Riemann invariants.* Physica D. 1993. V. 63.

P. 50–70.

[43] Ferapontov E. V. *On the matrix Hopf equation and integrable Hamiltonian systems of hydrodynamic type, which do not possess Riemann invariants.* Phys. Lett. A. 1993. V. 179. P. 391–397.

[44] Ferapontov E. V. *Dupin hypersurfaces and integrable Hamiltonian systems of hydrodynamic type which do not possess Riemann invariants.* Differential Geometry Appl. 1995. V. 5. No. 2. P. 121–152.

[45] Ferapontov E. V. *Several conjectures and results in the theory of integrable Hamiltonian systems of hydrodynamic type, which do not possess Riemann invariants.* Teoret. Mat. Fiz. 1994. V. 99. No. 2. P. 257–262; English transl. in Theoret. and Math. Phys. 1994. V. 99.

[46] Ferapontov E. V. *Nonlocal Hamiltonian operators of hydrodynamic type: differential geometry and applications.* In: Topics in Topology and Mathematical Physics. Ed. S. P. Novikov. Amer. Math. Soc., Providence, RI. 1995. P. 33–58.

[47] Ferapontov E. V. *Compatible Poisson brackets of hydrodynamic type.* J. Phys. A. 2001. V. 34. P. 2377–2388; arXiv: math.DG/0005221.

[48] Ferapontov E. V., Fordy A. P. *Separable Hamiltonians and integrable systems of hydrodynamic type.* J. Geom. Phys. 1997. V. 21. No. 2. P. 169–182.

[49] Ferapontov E. V., Fordy A. P. *Non-homogeneous systems of hydrodynamic type, related to quadratic Hamiltonians with electromagnetic term.* Physica D. 1997. V. 108. P. 350–364.

[50] Ferapontov E. V., Galvão C. A. P., Mokhov O. I., Nutku Y. *Bi-Hamiltonian structure of equations of associativity in 2D topological field theory.* Comm. Math. Phys. 1997. V. 186. P. 649–669.

[51] Ferapontov E. V., Mokhov O. I. *On the Hamiltonian representation of the associativity equations.* In: Algebraic aspects of integrable systems: In memory of Irene Dorfman. Eds. I. M. Gelfand, A. S. Fokas. Birkhäuser. Boston. 1996. P. 75–91.

[52] Ferapontov E. V., Mokhov O. I. *The equations of the associativity as hydrodynamic type systems: Hamiltonian representation and integrability.* In: Proceedings of the First International Workshop "Nonlinear Physics. Theory and Experiment. Nature, Structure and Properties of Nonlinear Phenomena". Le Sirenuse, Gallipoli (Lecce), Italy. 29 June–7 July, 1995. Eds. E. Alfinito, M. Boiti, L. Martina, F. Pempinelli. World Scientific Publishing. River Edge, NJ. 1996. P. 104–115.

[53] Ferapontov E. V., Pavlov M. V. *Quasiclassical limit of coupled KdV equations. Riemann invariants and multi-Hamiltonian structure.* Physica D. 1991. V. 52. P. 211–219.

[54] Flaschka H., Forest M. G., McLaughlin D. W. *Multiphase averaging and the inverse spectral solution of the Korteweg-de Vries equation.* Comm. Pure Appl. Math. 1980. V. 33. No. 6. P. 739–784.

[55] Fokas A. S., Fuchssteiner B. *On the structure of symplectic operators and hereditary symmetries.* Lettere al Nuovo Cimento. 1980. V. 28. No. 8. P. 299–303.

[56] Fordy A. P., Mokhov O. I. *On a special class of compatible Poisson structures of hydrodynamic type.* Physica D. 2001. V. 152–153. P. 475–490.

[57] Fordy A. P., Reyman A. G., Semenov-Tian-Shansky M. A. *Classical R-matrices and compatible Poisson brackets for coupled KdV systems.* Lett. Math. Phys. 1989. V. 17. No. 1. P. 25–29.

[58] Fuchssteiner B. *Application of hereditary symmetries to nonlinear evolution equations.* Nonlinear Analysis. Theory, Methods and Applications. 1979. V. 3. P. 849–862.

[59] Fuchssteiner B. *A symmetry approach to exactly solvable evolution equations.* J. Math. Phys. 1980. V. 21. P. 1318–1325.

[60] Fuchssteiner B., Fokas A. S. *Symplectic structures, their Bäcklund transformations and hereditary symmetries.* Physica D. 1981. V. 4. P. 47–66.

[61] Ganzha E. I. *Basic series of non-homogeneous systems of hydrodynamic type with constant matrices, which possess first-order conservation laws.* Fundamental and Applied Math. 1995. V. 1. No. 3. P. 641–648.

[62] Ganzha E. I. *On higher conservation laws of non-homogeneous systems of hydrodynamic type.* Vestnik Moskov. Univ. Ser. 1. 1996. No. 1. P. 86–87; English transl. in Moscow Univ. Math. Bull. 1996. V. 51. No. 1.

[63] Gardner C. S. *Korteweg–de Vries equation and generalizations. IV. The Korteweg–de Vries equation as a Hamiltonian system.* J. Math. Phys. 1971. V. 12. No. 8. P. 1548–1551.

[64] Gardner C. S., Greene J. M., Kruskal M. D., Miura R. M. *Korteweg–de Vries equation and generalizations. VI. Methods for exact solution.* Comm. Pure Appl. Math. 1974. V. 27. P. 97–133.

[65] Gelfand I. M., Dikii L. A. *Fractional powers of operators, and Hamiltonian systems.* Funktsional. Analiz i ego Prilozhen. 1976. V. 10. No. 4. P. 13–29; English transl. in Functional Anal. Appl. 1976. V. 10. No. 4.

[66] Gelfand I. M., Dikii L. A. *The resolvent, and Hamiltonian systems.* Funktsional. Analiz i ego Prilozhen. 1977. V. 11. No. 2.

P. 11–27; English transl. in Functional Anal. Appl. 1977. V. 11. No. 2.

[67] Gelfand I. M., Dorfman I. Ya. *Hamiltonian operators and algebraic structures associated with them.* Funktsional. Analiz i ego Prilozhen. 1979. V. 13. No. 4. P. 13–30; English transl. in Functional Anal. Appl. 1979. V. 13. No. 4. P. 246–262.

[68] Gelfand I. M., Dorfman I. Ya. *Schouten bracket and Hamiltonian operators.* Funktsional. Analiz i ego Prilozhen. 1980. V. 14. No. 3. P. 71–74; English transl. in Functional Anal. Appl. 1980. V. 14. No. 3.

[69] Gelfand I. M., Dorfman I. Ya. *Hamiltonian operators and infinite-dimensional Lie algebras.* Funktsional. Analiz i ego Prilozhen. 1981. V. 15. No. 3. P. 23–40; English transl. in Functional Anal. Appl. 1981. V. 15. No. 3. P. 173–187.

[70] Gelfand I. M., Dorfman I. Ya. *Hamiltonian operators and the classic Yang–Baxter equation.* Funktsional. Analiz i ego Prilozhen. 1982. V. 16. No. 4. P. 1–9; English transl. in Functional Anal. Appl. 1982. V. 16. No. 4.

[71] Gelfand I. M., Manin Yu. I., Shubin M. A. *Poisson brackets and the kernel of the variational derivative in the formal calculus of variations.* Funktsional. Analiz i ego Prilozhen. 1976. V. 10. No. 4. P. 30–34; English transl. in Functional Anal. Appl. 1976. V. 10. No. 4.

[72] Gibbons J., Tsarev S. P. *Reductions of the Benney equations.* Phys. Lett. A. 1996. V. 211. No. 1. P. 19–24.

[73] Gibbons J., Tsarev S. P. *Conformal maps and reductions of the Benney equations.* Phys. Lett. A. 1999. V. 258. P. 263–271.

[74] Gorsky A., Marshakov A., Orlov A., Rubtsov V. *On the third Poisson structure of the KdV equation.* Teoret. Mat. Fiz. 1995. V. 103. No. 3. P. 461–466; English transl. in Theoret. Math. Phys. 1995. V. 103. No. 3. P. 701–706.

[75] Grinberg N. I. *On Poisson brackets of hydrodynamic type with a degenerate metric.* Uspekhi Mat. Nauk. 1985. V. 40. No. 4.

P. 217–218; English transl. in Russian Math. Surveys. 1985.
V. 40. No. 4. P. 231–232.

[76] Haantjes A. *On X_{n-1}-forming sets of eigenvectors.* Indagationes Mathematicae. 1955. V. 17. No. 2. P. 158–162.

[77] Kac V. *Infinite dimensional Lie algebras.* Cambridge. Cambridge University Press. 1990; Russian transl.: Mir. Moscow. 1993.

[78] Kontsevich M. *Homological algebra of mirror symmetry.* In: Proceedings of the International Congress of Mathematicians. August 3–11, 1994, Zürich, Switzerland. V. 1. Birkhäuser. Basel. 1995. P. 120–139.

[79] Kontsevich M., Manin Yu. I. *Gromov–Witten classes, quantum cohomology, and enumerative geometry.* Comm. Math. Phys. 1994. V. 164. P. 525–562.

[80] Kosmann-Schwarzbach Y., Magri F. *Poisson–Nijenhuis structures.* Annales de l'Institut Henri Poincaré, série A (Physique théorique). 1990. V. 53. No. 1. P. 35–81.

[81] Krichever I. M., Novikov S. P. *Holomorphic bundles and nonlinear equations. Finite-gap solutions of rank 2.* Dokl. Akad. Nauk SSSR. 1979. V. 247. No. 1. P. 33–37; English transl. in Soviet Math. Dokl. 1979. V. 20.

[82] Kupershmidt B. A. *Invariance of the stationary manifold of a conserved density.* Phys. Lett. A. 1987. V. 121. Nos. 8, 9. P. 395–398.

[83] Kupershmidt B. A. *GL_2-orbit of the Korteweg–de Vries equation.* Phys. Lett. A. 1991. V. 156. Nos. 1, 2. P. 53–60.

[84] Kupershmidt B. A., Wilson G. *Modifying Lax equations and the second hamiltonian structure.* Invent. Math. 1981. V. 62. No. 3. P. 403–436.

[85] Lax P. D. *Periodic solutions of the KdV equation.* Lect. Appl. Math. 1974. V. 15. P. 85–96.

[86] Lax P. D. *Almost periodic solutions of the KdV equation.* SIAM Review. 1976. V. 18. No. 3. P. 351–375.

[87] Luke J. C. *A perturbation method for nonlinear dispersive wave problems.* Proc. Royal Soc. London. Ser. A. 1966. V. 292. No. 1430. P. 403–412.

[88] Magri F. *A simple model of the integrable Hamiltonian equation.* J. Math. Phys. 1978. V. 19. No. 5. P. 1156–1162.

[89] Magri F. *A geometrical approach to the nonlinear solvable equations.* In: Nonlinear Evolution Equations and Dynamical Systems. Eds. M. Boiti, F. Pempinelli, G. Soliani. Lecture Notes in Physics. 1980. V. 120. Springer-Verlag. New York. P. 233–263.

[90] Magri F., Morosi C., Tondo G. *Nijenhuis G-manifolds and Lenard bicomplexes.* Comm. Math. Phys. 1988. V. 115. P. 457–475.

[91] Maltsev A. Ya. *The conservation of the Hamiltonian structures in Whitham's method of averaging.* Izvestiya, Mathematics. 1999. V. 63. No. 6. P. 1171–1201; solv-int/9611008.

[92] Maltsev A. Ya. *The averaging of local field-theoretical Poisson brackets.* Uspekhi Mat. Nauk. 1997. V. 52. No. 2. P. 177–178; English transl. in Russian Math. Surveys. 1997. V. 52. No. 2. P. 409–411.

[93] Maltsev A. Ya. *The non-local Poisson brackets and the Whitham method.* Uspekhi Mat. Nauk. 1999. V. 54. No. 6; English transl. in Russian Math. Surveys. 1999. V. 54. No. 6. P. 1252–1253.

[94] Maltsev A. Ya. *The averaging of non-local Hamiltonian structures in Whitham's method.* Preprint SISSA-9/2000/FM; solv-int/9910011.

[95] Maltsev A. Ya., Novikov S. P. *On the local systems hamiltonian in the weakly nonlocal Poisson brackets.* Physica D. 2001. V. 156. No. 1–2. P. 53–80; arXiv:nlin.SI/0006030.

[96] Maltsev A. Ya., Pavlov M. V. *On Whitham's averaging method.* Funktsional. Analiz i ego Prilozhen. 1995. V. 29. No. 1. P. 7–24; English transl. in Functional Anal. Appl. 1995. V. 29. No. 1.

[97] Manin Yu. I. *Algebraic aspects of non-linear differential equations.* In: Itogi Nauki i Tekhniki. Sovremennye Problemy Matem. 1978. V. 11. Moscow. VINITI. P. 5–152; English transl. in J. Soviet Math. 1979. V. 11. No. 1.

[98] Miura R. M. *Korteweg–de Vries equation and generalizations. I. A remarkable explicit non-linear transformation.* J. Math. Phys. 1968. V. 9. No. 8. P. 1202–1204.

[99] Meshkov A. G. *Hamiltonian and recursion operators for two-dimensional scalar fields.* Phys. Lett. A. 1992. V. 170. No. 6. P. 405–408.

[100] Mokhov O. I. *The Hamiltonian property of an evolutionary flow on the set of stationary points of its integral.* Uspekhi Mat. Nauk. 1984. V. 39. No. 4. P. 173–174; English transl. in Russian Math. Surveys. 1984. V. 39. No. 4. P. 133–134.

[101] Mokhov O. I. *Local third-order Poisson brackets.* Uspekhi Mat. Nauk. 1985. V. 40. No. 5. P. 257–258; English transl. in Russian Math. Surveys. 1985. V. 40. No. 5. P. 233–234.

[102] Mokhov O. I. *Geometry of commuting differential operators of rank 3 and Hamiltonian flows.* Ph.D. Dissertation (Phys. and Math.). Moscow State University. Moscow. 1984 [in Russian].

[103] Mokhov O. I. *Hamiltonian differential operators and contact geometry.* Funktsional. Analiz i ego Prilozhen. 1987. V. 21. No. 3. P. 53–60; English transl. in Functional Anal. Appl. 1987. V. 21. No. 3. P. 217–223.

[104] Mokhov O. I. *On the Hamiltonian property of an arbitrary evolution system on the set of stationary points of its integral.* Izvestiya Akad. Nauk SSSR. Ser. Matem. 1987. V. 51. No. 6. P. 53–60; English transl. in Math. USSR – Izvestiya. 1988. V. 31. No. 3. P. 657–664.

[105] Mokhov O. I. *On Poisson brackets of Dubrovin–Novikov type (DN-brackets)*. Funktsional. Anal. i ego Prilozhen. 1988. V. 22. No. 4. P. 92–93; English transl. in Functional Anal. Appl. 1988. V. 22. No. 4. P. 336–338.

[106] Mokhov O. I. *Canonical variables for vortex two-dimensional hydrodynamics of an incompressible fluid.* Teoret. Mat. Fiz. 1989. V. 78. No. 1. P. 136–139; English transl. in Theoret. Math. Phys. 1989. V. 78. No. 1. P. 97–99.

[107] Mokhov O. I. *Contact geometry and calculus of variations.* In: Geometry, Topology, and Applications. Moscow. MIP. 1990. P. 111–115 [in Russian].

[108] Mokhov O. I. *A Hamiltonian structure of evolution with respect to the space variable x for the Korteweg–de Vries equation.* Uspekhi Mat. Nauk. 1990. V. 45. No. 1. P. 181–182; English transl. in Russian Math. Surveys. 1990. V. 45. No. 1. P. 218–220.

[109] Mokhov O. I. *Symplectic forms on loop space and Riemannian geometry.* Funktsional. Analiz i ego Prilozhen. 1990. V. 24. No. 3. P. 86–87; English transl. in Functional Anal. Appl. 1990. V. 24. No. 3. P. 247–249.

[110] Mokhov O. I. *Homogeneous symplectic structures of second order on loop spaces and symplectic connections.* Funktsional. Analiz i ego Prilozhen. 1991. V. 25. No. 2. P. 65–67; English transl. in Functional Anal. Appl. 1991. V. 25. No. 2. P. 136–137.

[111] Mokhov O. I. *Canonical Hamiltonian representation of the Krichever–Novikov equation.* Matem. Zametki. 1991. V. 50. No. 3. P. 87–96; English transl. in Math. Notes. 1991. V. 50. No. 3. P. 939–945.

[112] Mokhov O. I. *Theorem on the Hamiltonian property of an arbitrary evolution flow on the set of stationary points of its integral and Noether theorem.* In: Asymptotical methods for solving differential equations. Institute of Mathematics, Ural Branch of the Russian Acad. Sci. Ufa. 1992. P. 74–75 [in Russian].

[113] Mokhov O.I. *Two-dimensional non-linear σ-models in field theory: symplectic approach.* In: Abstracts of 9th Conference "Modern Group Analysis. Methods and Applications". 24–30 June, 1992. Scientific and Research Radiophysical Institute. Nizhnii Novgorod. 1992. P. 38 [in Russian].

[114] Mokhov O.I. *On complexes of homogeneous forms on loop spaces of smooth manifolds and their cohomology groups.* Uspekhi Mat. Nauk. 1996. V. 51. No. 2. P. 141–142; English transl. in Russian Math. Surveys. 1996. V. 51. No. 2. P. 341–342.

[115] Mokhov O.I. *Symplectic and Poisson structures on loop spaces of smooth manifolds, and integrable systems.* D.Sc. Dissertation (Phys. and Math.). Steklov Mathematical Institute. Moscow. 1996 [in Russian].

[116] Mokhov O.I. *On cohomology groups of complexes of homogeneous forms on loop spaces of smooth manifolds.* Funktsional. Analiz i ego Prilozhen. 1998. V. 32. No. 3. P. 22–34; English transl. in Functional Anal. Appl. 1998. V. 32. No. 3. P. 162–171.

[117] Mokhov O.I. *Differential geometry of symplectic and Poisson structures on loop spaces of smooth manifolds, and integrable systems.* In: Loop spaces and groups of diffeomorphisms. Eds. A.G. Sergeev. Trudy Matem. Inst. Akad. Nauk. 1997. V. 217. Moscow. Nauka. P. 100–134; English transl. in Proceedings of Steklov Mathematical Institute (Moscow). 1997. V. 217. P. 91–125.

[118] Mokhov O.I. *On compatible Poisson structures of hydrodynamic type.* Uspekhi Mat. Nauk. 1997. V. 52. No. 6. P. 171–172; English transl. in Russian Math. Surveys. 1997. V. 52. No. 6. P. 1310–1311.

[119] Mokhov O.I. *On compatible potential deformations of Frobenius algebras and associativity equations.* Uspekhi Mat. Nauk. 1998. V. 53. No. 2. P. 153–154; English transl. in Russian Math. Surveys. 1998. V. 53. No. 2. P. 396–397.

210   *O.I. MOKHOV*

[120] Mokhov O. I. *Vorticity equation of two-dimensional hydrodynamics of an incompressible fluid as canonical Hamiltonian system.* Phys. Lett. A. 1989. V. 139. No. 8. P. 363–368.

[121] Mokhov O. I. *Hamiltonian systems of hydrodynamic type and constant curvature metrics.* Phys. Lett. A. 1992. V. 166. Nos. 3, 4. P. 215–216.

[122] Mokhov O. I. *On the canonical variables for two-dimensional vortex hydrodynamics of incompressible fluid.* In: International Series of Numerical Mathematics. 1992. V. 106. Birkhäuser. Basel. P. 215–221.

[123] Mokhov O. I. *Two-dimensional nonlinear sigma models and symplectic geometry on loop spaces of (pseudo)-Riemannian manifolds.* In: Nonlinear evolution equations and dynamical systems. Proceedings of the 8th International Workshop (NEEDS'92), 6–17 July, 1992, Dubna, Russia. Ed. V. G. Makhan'kov. World Scientific Publishing. Singapore. 1993. P. 444–456; hep-th/9301048.

[124] Mokhov O. I. *Symplectic and Poisson geometry on loop spaces of manifolds and nonlinear equations.* In: Topics in Topology and Mathematical Physics. Ed. S. P. Novikov. Amer. Math. Soc., Providence, RI. 1995. P. 121–151; arXiv: hep-th/9503076.

[125] Mokhov O. I. *Differential equations of associativity in 2D topological field theories and geometry of nondiagonalizable systems of hydrodynamic type.* In: Internat. Conference on Integrable Systems "Nonlinearity and Integrability: from Mathematics to Physics", February 21–24, 1995, Montpellier, France.

[126] Mokhov O. I. *Poisson and symplectic geometry on loop spaces of smooth manifolds.* In: Geometry from the Pacific Rim. Proceedings of the Pacific Rim Geometry Conference held at National University of Singapore, Republic of Singapore, December 12–17, 1994. Eds. A. J. Berrick, B. Loo, H.-Y. Wang. Walter de Gruyter. Berlin. 1997. P. 285–309.

[127] Mokhov O.I. *Symplectic and Poisson structures on loop spaces of smooth manifolds, and integrable systems.* Uspekhi Mat. Nauk. 1998. V. 53. No. 3. P. 85–192; English transl. in Russian Math. Surveys. 1998. V. 53. No. 3. P. 515–622.

[128] Mokhov O.I. *On the equations of associativity and compatible Poisson structures of hydrodynamic type.* The International Conference "Nonlinear Systems, Solitons and Geometry". 19 – 25.10.1997. Mathematisches Forschungsinstitut Oberwolfach. Tagungsbericht 40/1997. P. 14.

[129] Mokhov O.I. *Compatible Poisson structures of hydrodynamic type and the equations of associativity in two-dimensional topological field theory.* Reports on Mathematical Physics. 1999. V. 43. No. 1/2. P. 247–256.

[130] Mokhov O.I. *Compatible Poisson structures of hydrodynamic type and associativity equations.* Trudy Matem. Inst. Akad. Nauk. 1999. V. 225. Moscow. Nauka. P. 284–300; English transl. in Proceedings of the Steklov Institute of Mathematics (Moscow). 1999. V. 225. P. 269–284.

[131] Mokhov O.I. *Compatible and almost compatible metrics.* Uspekhi Mat. Nauk. 2000. V. 55. No. 4. P. 217–218; English transl. in Russian Math. Surveys. 2000. V. 55. No. 4. P. 819–821.

[132] Mokhov O.I. *Compatible and almost compatible pseudo-Riemannian metrics.* Funktsional. Analiz i ego Prilozhen. 2001. V. 35. No. 2. P. 24–36; English transl. in Functional Anal. Appl. 2001. V. 35. No. 2. P. 100-110; arXiv: math.DG/0005051.

[133] Mokhov O.I. *Compatible flat metrics.* International Congress on Differential Geometry in Memory of Alfred Gray. 18 – 23 September 2000. Bilbao, Spain. The University of the Basque Country. 2000. P. 69–70.

[134] Mokhov O.I. *Flat pencils of metrics and integrable reductions of Lamé's equations.* Uspekhi Mat. Nauk. 2001. V. 56. No. 2. P. 221–222; English transl. in Russian Math. Surveys. 2001. V. 56. No. 2. P. 416–418.

212   *O.I. MOKHOV*

[135] Mokhov O. I. *Flat pencils of metrics and integrable reductions of the Lamé equations.* International Conference "Differential Equations and Related Topics", dedicated to the Centenary Anniversary of Ivan G. Petrovskii (1901–1973). 22 – 27 May 2001. Moscow. Moscow University Press. 2001. P. 277.

[136] Mokhov O. I. *Integrability of the equations for nonsingular pairs of compatible flat metrics.* Teoret. Mat. Fiz. 2002. V. 130. No. 2. P. 233–250; English transl. in Theoret. Math. Phys. 2002. V. 130. No. 2. P. 198–212; arXiv: math.DG/0005081 (2000).

[137] Mokhov O. I. *Compatible flat metrics.* Journal of Applied Math. 2002. V. 2. No. 7. P. 337–370; arXiv: math.DG/0201224 (2002).

[138] Mokhov O. I. *Compatible Dubrovin–Novikov Hamiltonian operators and the Lie derivative.* Uspekhi Mat. Nauk. 2001. V. 56. No. 6. P. 161–162; English transl. in Russian Math. Surveys. 2001. V. 56. No. 6. P. 1175–1176.

[139] Mokhov O. I. *Integrable bi-Hamiltonian systems of hydrodynamic type.* Uspekhi Mat. Nauk. 2002. V. 57. No. 1. P. 157–158; English transl. in Russian Math. Surveys. 2002. V. 57. No. 1. P. 153–154.

[140] Mokhov O. I. *Compatible Dubrovin–Novikov Hamiltonian operators, Lie derivative, and integrable systems of hydrodynamic type.* Proceedings of the International Conference "Nonlinear Evolution Equations and Dynamical Systems," Cambridge (England), July 24–30, 2001. Teoret. Mat. Fiz. 2002. V. 133. No. 2. P. 279–288; English transl. in Theoret. Math. Phys. 2002. V. 133. No. 2. P. 1555–1562; arXiv: math.DG/0201281 (2002).

[141] Mokhov O. I. *The Lax pair for non-singular pencils of metrics of constant Riemannian curvature.* Uspekhi Mat. Nauk. 2002. V. 57. No. 3. P. 155–156; English transl. in Russian Math. Surveys. 2002. V. 57. No. 3. P. 603–605.

[142] Mokhov O. I. *Compatible metrics of constant Riemannian curvature: local geometry, nonlinear equations, and integra-*

bility. Funktsional. Analiz i ego Prilozhen. 2002. V. 36. No. 3. P. 36–47; English transl. in Functional Anal. Appl. 2002. V. 36. No. 3. P. 196–204; arXiv: math.DG/0201280 (2002).

[143] Mokhov O. I. *Integrable bi-Hamiltonian hierarchies generated by compatible metrics of constant Riemannian curvature.* Uspekhi Mat. Nauk. 2002. V. 57. No. 5. P. 157–158; English transl. in Russian Math. Surveys. 2002. V. 57. No. 5. P. 999–1001.

[144] Mokhov O. I. *Liouville canonical form for compatible nonlocal Poisson brackets of hydrodynamic type, and integrable hierarchies.* Funktsional. Analiz i ego Prilozhen. 2003. V. 37. No. 2. P. 28–40; English transl. in Functional Anal. Appl. 2003. V. 37. No. 2. P. 103–113; arXiv: math.DG/0201223 (2002).

[145] Mokhov O. I. *Compatible nonlocal Poisson brackets of hydrodynamic type and integrable hierarchies related to them.* Teoret. Mat. Fiz. 2002. V. 132. No. 1. P. 60–73; English transl. in Theoret. Math. Phys. 2002. V. 132. No. 1. P. 908–916; arXiv: math.DG/0201242 (2002).

[146] Mokhov O. I. *The Lax pairs for compatible non-local Hamiltonian operators of hydrodynamic type.* Uspekhi Mat. Nauk. 2002. V. 57. No. 6. P. 189–190; English transl. in Russian Math. Surveys. 2002. V. 57. No. 6. P. 1234–1235.

[147] Mokhov O. I. *Compatible nonlocal Poisson brackets of hydrodynamic type and integrable reductions of the Lamé equations.* Proceedings of the Workshop "Nonlinear Physics: Theory and Experiment. II." Gallipoli, Italy. 27 June – 6 July 2002. World Scientific Publishing. New Jersey–London–Singapore–Hong Kong. 2003. P. 200–206.

[148] Mokhov O. I. *Compatible nonlocal Poisson brackets of hydrodynamic type: local Riemannian geometry and integrability.* International Conference "Kolmogorov and Contemporary Mathematics" in commemoration of the centennial of Andrei Nikolaevich Kolmogorov (1903–1987). 16 – 21 June 2003. Moscow. Moscow State University. 2003. P. 827–828.

[149] Mokhov O.I. *Lax pairs for the equations describing compatible non-local Poisson brackets of hydrodynamic type, and integrable reductions of the Lamé equations.* Teoret. Mat. Fiz. 2004. V. 138. No. 2. P. 283–296; English transl. in Theoret. Math. Phys. 2004. V. 138. No. 2; arXiv: math.DG/0202036 (2002).

[150] Mokhov O.I. *Quasi-Frobenius algebras and their integrable N-parameter deformations generated by compatible (N × N) metrics of constant Riemannian curvature.* Teoret. Mat. Fiz. 2003. V. 136. No. 1. P. 20–29; English transl. in Theoret. Math. Phys. 2003. V. 136. No. 1. P. 908–916; arXiv: math.DG/0209262 (2002).

[151] Mokhov O.I., Ferapontov E.V. *Non-local Hamiltonian operators of hydrodynamic type related to metrics of constant curvature.* Uspekhi Mat. Nauk. 1990. V. 45. No. 3. P. 191–192; English transl. in Russian Math. Surveys. 1990. V. 45. No. 3. P. 218–219.

[152] Mokhov O.I., Ferapontov E.V. *Hamiltonian pairs associated with skew-symmetric Killing tensors on spaces of constant curvature.* Funktsional. Analiz i ego Prilozhen. 1994. V. 28. No. 2. P. 60–63; English transl. in Functional Anal. Appl. 1994. V. 28. No. 2. P. 123–125.

[153] Mokhov O.I., Ferapontov E.V. *Equations of associativity in two-dimensional topological field theory as integrable Hamiltonian non-diagonalizable systems of hydrodynamic type.* Funktsional. Analiz i ego Prilozhen. 1996. V. 30. No. 3. P. 62–72; English transl. in Functional Anal. Appl. 1996. V. 30. No. 3. P. 195–203; hep-th/9505180.

[154] Mokhov O.I., Nutku Y. *Homogeneous Poisson brackets of Dubrovin–Novikov type and their nonlocal generalizations.* Preprint. TÜBİTAK – Marmara Research Center, Research Institute for Basic Sciences, Gebze, Turkey. 1996.

[155] Mokhov O.I., Nutku Y. *Bianchi transformation between the real hyperbolic Monge–Ampère equation and the Born–Infeld equation.* Lett. Math. Phys. 1994. V. 32. No. 2. P. 121–123.

[156] Nijenhuis A. X_{n-1}-forming sets of eigenvectors. Indagationes Mathematicae. 1951. V. 13. No. 2. P. 200–212.

[157] Nijenhuis A. Geometric aspects of formal differential operators on tensor fields. In: Proceedings of the International Congress of Mathematicians. 1958. Edinburgh. P. 463–469.

[158] Novikov S. P. Periodic problem for the Korteweg–de Vries equation. Funktsional. Analiz i ego Prilozhen. 1974. V. 8. No. 3. P. 54–66; English transl. in Functional Anal. Appl. 1974. V. 8. No. 3. P. 236–246.

[159] Novikov S. P. Two-dimensional Schrödinger operators in periodic fields. In: Itogi Nauki i Tekhniki. Sovremennye Problemy Matem. 1983. V. 23. Moscow. VINITI. P. 3–32; English transl. in J. Soviet Math. 1985. V. 28. No. 1.

[160] Novikov S. P. The geometry of conservative systems of hydrodynamic type. The method of averaging for field-theoretical systems. Uspekhi Mat. Nauk. 1985. V. 40. No. 4. P. 79–89; English transl. in Russian Math. Surveys. V. 40. No. 4. P. 85–98.

[161] Novikov S. P. Differential geometry and hydrodynamics of soliton lattices. In: Important developments in soliton theory. Eds. A. S. Fokas, V. E. Zakharov. Springer Ser. in Nonlinear Dyn. Berlin. Springer-Verlag. 1993. P. 242–256.

[162] Novikov S. P. (ed.) Theory of solitons: Inverse problem method. Moscow. Nauka. 1980; English transl.: Plenum Press. New York. 1984.

[163] Novikov S. P., Maltsev A. Ya. The Liouville form of averaged Poisson brackets. Uspekhi Mat. Nauk. 1993. V. 48. No. 1. P. 155–156; English transl. in Russian Math. Surveys. 1993. V. 48. No. 1.

[164] Nutku Y. Canonical formulation of shallow water waves. J. Phys. A. 1983. V. 16. P. 4195–4201.

[165] Nutku Y. Hamiltonian formulation of the KdV equation. J. Math. Phys. 1984. V. 25. No. 6. P. 2007–2008.

[166] Nutku Y. *Hamiltonian structure of real Monge–Ampère equations.* J. Phys. A.: Math. Gen. 1996. V. 29. P. 3257–3280.

[167] Nutku Y., Sarıoğlu O. *An integrable family of Monge–Ampère equations and their multi-Hamiltonian structure.* Phys. Lett. A. 1993. V. 173. No. 3. P. 270–274.

[168] Olver P. *Applications of Lie groups to differential equations.* New York. Springer-Verlag. 1986; Russian transl.: Mir. Moscow. 1989.

[169] Pavlov M. V. *Multi-Hamiltonian structures of the Whitham equations.* Dokl. Akad. Nauk SSSR. 1994. V. 338. No. 2. P. 165–167; English transl. in Soviet Math. Dokl. 1995. V. 50.

[170] Pavlov M. V. *Elliptic coordinates and multi-Hamiltonian structures of systems of hydrodynamic type.* Dokl. Akad. Nauk SSSR. 1994. V. 339. No. 1. P. 21–23; English transl. in Soviet Math. Dokl. 1995. V. 50.

[171] Pavlov M. V., Tsarev S. P. *On conservation laws of Benney equations.* Uspekhi Mat. Nauk. 1991. V. 46. No. 4. P. 169–170; English transl. in Russian Math. Surveys. 1991. V. 46. No. 4.

[172] Potemin G. V. *Some problems of differential geometry and algebraic geometry in soliton theory.* Ph.D. Dissertation (Phys. and Math.). Moscow State University. Moscow. 1991 [in Russian].

[173] Potemin G. V. *On Poisson brackets of differential-geometric type.* Dokl. Akad. Nauk SSSR. 1986. V. 286. No. 1. P. 39–42; English transl. in Soviet Math. Dokl. 1986. V. 33. P. 30–33.

[174] Reeb G. *Quelques propriétés globales des trajectoires de la dynamique dues à l'existence de l'invariant intégral de M. Élie Cartan.* Comptes Rendus Acad. Sci. Paris. 1949. V. 229. No. 20. P. 969–971.

[175] Reeb G. *Variétés de Riemann dont toutes les géodésiques sont fermées.* Bull. Cl. Sciences. Acad. Royale Belgique. 5 Série. 1950. V. 36. No. 4. P. 324–329.

[176] Ruan Y., Tian G. *A mathematical theory of quantum cohomology.* J. Differential Geometry. 1995. V. 42. No. 2. P. 259–367.

[177] Sokolov V. V. *On the Hamiltonian property of the Krichever–Novikov equation.* Dokl. Akad. Nauk SSSR. 1984. V. 277. No. 1. P. 48–50; English transl. in Soviet Math. Dokl. 1984. V. 30.

[178] Tsarev S. P. *On Poisson brackets and one-dimensional Hamiltonian systems of hydrodynamic type.* Dokl. Akad. Nauk SSSR. 1985. V. 282. No. 3. P. 534–537; English transl. in Soviet Math. Dokl. 1985. V. 31.

[179] Tsarev S. P. *The Hamiltonian property of stationary and inverted equations of the mechanics of continuous media and mathematical physics.* Matem. Zametki. 1989. V. 46. No. 1. P. 105–111; English transl. in Math. Notes. 1989. V. 46. Nos. 1-2. P. 569–573.

[180] Tsarev S. P. *Differential-geometric methods for integrating systems of hydrodynamic type.* D.Sc. Dissertation (Phys. and Math.). Steklov Mathematical Institute. Moscow. 1993 [in Russian].

[181] Tsarev S. P. *Geometry of Hamiltonian systems of hydrodynamic type. The generalized hodograph method.* Izvestiya Akad. Nauk SSSR. Ser. Matem. 1990. V. 54. No. 5. P. 1048–1068; English transl. in Math. USSR – Izvestiya. 1990. V. 54. No. 5. P. 397–419.

[182] Veselov A. P. *Hamiltonian formalism for Novikov–Krichever equations on the commutativity of two operators.* Funktsional. Analiz i ego Prilozhen. 1979. V. 13. No. 1. P. 1–7; English transl. in Functional Anal. Appl. 1979. V. 13. No. 1.

[183] Veselov A. P. *Integrable systems with discrete time and difference operators.* Funktsional. Analiz i ego Prilozhen. 1988. V. 22. No. 2. P. 1–13; English transl. in Functional Anal. Appl. 1988. V. 22. No. 2. P. 83–93.

[184] Veselov A. P. *What is an integrable map?* In: Integrability and kinetic equations for solitons. Kiev. Naukova Dumka. 1990. P. 44–64 [in Russian].

[185] Veselov A. P. *Integrable maps*. Uspekhi Mat. Nauk. 1991. V. 46. No. 5. P. 3–45; English transl. in Russian Math. Surveys. 1991. V. 46. No. 5. P. 1–51.

[186] Vinogradov A. M. *Hamiltonian structures in field theory*. Dokl. Akad. Nauk SSSR. 1978. V. 241. No. 1. P. 18–21; English transl. in Soviet Math. Dokl. 1978. V. 19.

[187] Weinstein A. *On the volume of manifolds all of whose geodesics are closed*. J. Differential Geom. 1974. V. 9. No. 4. P. 513–517.

[188] Whitham G. *Linear and nonlinear waves*. Wiley. New York. 1974.

[189] Witten E. *On the structure of the topological phase of two-dimensional gravity*. Nuclear Phys. B. 1990. V. 340. P. 281–332.

[190] Witten E. *Two-dimensional gravity and intersection theory on moduli space*. Surveys in Diff. Geometry. 1991. V. 1. P. 243–310.

[191] Yano K. *On harmonic and Killing vector fields*. Ann. of Math. (2). 1952. V. 55. No. 1. P. 38–45.

[192] Yano K. *Some remarks on tensor fields and curvature*. Ann. of Math. (2). 1952. V. 55. No. 2. P. 328–347.

[193] Yano K., Bochner S. *Curvature and Betti numbers*. Princeton. Princeton University Press. 1953.

[194] Zakharov V. E. *Benney's equations and the quasi-classical approximation in the inverse problem method*. Funktsional. Analiz i ego Prilozhen. 1980. V. 14. No. 2. P. 15–24; English transl. in Functional Anal. Appl. 1980. V. 14. No. 2. P. 89–98.

[195] Zakharov V. E. *Description of the n-orthogonal curvilinear coordinate systems and Hamiltonian integrable systems of hydrodynamic type, I: Integration of the Lamé equations*. Duke Math. J. 1998. V. 94. No. 1. P. 103–139.

[196] Zakharov V. E., Faddeev L. D. *The Korteweg–de Vries equation is a completely integrable system.* Funktsional. Analiz i ego Prilozhen. 1971. V. 5. No. 4. P. 18–27; English transl. in Functional Anal. Appl. 1971. V. 5. No. 4. P. 280–287.

Index